THE LIBRARY
ST. MARY'S COLLEGE OF MARYLAND
ST. MARY'S CITY, MARYLAND 20686

084094

HISTORY, PHILOSOPHY AND SOCIOLOGY OF SCIENCE

Classics, Staples and Precursors

HISTORY, PHILOSOPHY AND SOCIOLOGY OF SCIENCE

Classics, Staples and Precursors

Selected By

YEHUDA ELKANA
ROBERT K. MERTON
ARNOLD THACKRAY
HARRIET ZUCKERMAN

THE ROLE OF SCIENTIFIC SOCIETIES IN THE SEVENTEENTH CENTURY

Martha Ornstein [Bronfenbrenner]

ARNO PRESS
A New York Times Company
New York – 1975

Reprint Edition 1975 by Arno Press Inc.

Reprinted from a copy in
  The Newark Public Library

HISTORY, PHILOSOPHY AND SOCIOLOGY OF SCIENCE:
Classics, Staples and Precursors
ISBN for complete set: 0-405-06575-2
See last pages of this volume for titles.

Manufactured in the United States of America

Library of Congress Cataloging in Publication Data

Ornstein, Martha, 1879-1915.
  The role of scientific societies in the seventeenth century.

  (History, philosophy, and sociology of science)
  Reprint of the ed. published by the University of Chicago Press, Chicago, which was originally presented as the author's thesis, Columbia, 1913.
  Bibliography: p.
  1. Science--History. 2. Scientific societies--History. I. Title. II. Series.
  Q125.08   1975   506'.2   74-26282
  ISBN 0-405-06609-0

THE RÔLE OF SCIENTIFIC SOCIETIES IN THE SEVENTEENTH CENTURY

By MARTHA ORNSTEIN

THE UNIVERSITY OF CHICAGO PRESS
CHICAGO · ILLINOIS

COPYRIGHT 1928 BY THE UNIVERSITY OF CHICAGO. COMPOSED,
PRINTED, AND PUBLISHED BY THE UNIVERSITY OF CHICAGO PRESS
IN MAY 1928

IN MEMORY OF
MARTHA ORNSTEIN BRONFENBRENNER
IDEALIST, TEACHER OF YOUTH,
SEEKER AFTER KNOWLEDGE

GRATEFUL AND DEVOTED FRIENDS HAVE
AFFECTIONATELY REPUBLISHED
THIS BOOK

EDITOR'S NOTE

In reprinting this study the Editor has endeavored not to alter the text as printed in 1913. He has almost without exception retained the quotations in the form found in the original publication.

MARTHA ORNSTEIN BRONFENBRENNER

Martha Ornstein Bronfenbrenner was born in Vienna, Austria, on August 19, 1878. She spent all her school years in Vienna, in the grammar school, high school, and then for two years in a Gymnasium, which roughly is the equivalent of our Freshman and Sophomore years at college. When she came to this country, in September, 1895, her preliminary education was therefore well advanced, and her main difficulty consisted in becoming sufficiently proficient in the use of the English language to enter Barnard College. To do this immediately became her goal and ambition. She promptly showed the ability and industry which were outstanding characteristics throughout her life by passing brilliant entrance examinations for Barnard in 1896. She completed the undergraduate course in three years instead of the usual four, although during the greater part of this period she supported herself by tutoring. She had a remarkable gift not only for acquiring but for imparting knowledge, so that her success as a teacher soon was recognized to be exceptional. In 1899 she received the A.B. degree, a Master's degree in 1900, and was elected a member of the Phi Beta Kappa Society when that organization was established at Barnard in 1901; in 1913 she was given her Ph.D. degree. For a number of years she specialized in mathematics, but came to a point when she felt that she had reached an *impasse* and could advance no farther. It was then that she turned to history, and her achievement in this field is attested by the Doctor's dissertation which now is being republished.

Notwithstanding Martha Ornstein's scholarly interests and attainments, she was a thoroughly well-rounded woman, human and sincere, with a friendliness which made of every pupil a devoted friend, in spite of the fact that she was a severe taskmaster. During all the years that she was working for higher degrees, she taught assiduously, summer as well as winter, and won the reputation of never having a pupil fail of admission to college.

In September, 1913, she married Dr. Jacob Bronfenbrenner and went to Pittsburgh to live. That winter she devoted herself to domestic life with as much zest as previously she had given to teaching and studying, and thoroughly enjoyed the duties of a housewife. Her son, Martin, was born in December, 1914, and her cup of happiness was full. Unfortunately she was granted only a few months of this new joy, for she died as a result of a motor accident in April, 1915. Her ability and personality are indicated by the fact that now, twelve years after her death, her friends are presenting this testimonial of their abiding affection and admiration.

FOREWORD

The distinctive trait of man is his power of accumulating knowledge, ranging from the practical information upon which we operate daily to mystic revelations and to metaphysical pronouncements on the dark nature of knowing itself. The confidant knowledge of one age may become the myth and superstition of the next. Interests change and assurances wax and wane. Theology was the queen of sciences in the Middle Ages but in the early part of the seventeenth century a feeble rival to the throne appeared who aspired to conquer a far more restricted realm, slowly but permanently. This new aspirant was experimental science whose retainers proclaimed the policy of pushing out their boundaries persistently but with the utmost caution. This was to result in what Bacon calls "the kingdom of man," in which knowledge acquired with the most scrupulous precautions against mistakes would continue to increase indefinitely, meanwhile ever bettering man's estate.

The respectable professors in the renowned seats of learning had no disposition to disfigure their academic gowns with splashes of acid, to fuss with childish devices in order to see how fast a ball would run down an inclined plane, or to be caught scooping up pond water covered with a disgusting scum. They had comfortable and long-revered works by ancient authorities which they could examine with dignity and expound to their classes without messing their pinafores, or distressing their gouty limbs. What was good enough for Aristotle and Pliny was good enough for them.

The present writer, impressed with the incredible revolution produced in man's ways of thinking and doing by the scientific discoveries of the past three centuries, found himself asking what part the ancient and honorable centers of learning—the universities—had had in the advancement of knowledge. There may have been a trace of malice aforethought in the query—some foresight of that long withheld work on *The Higher Learning* by his friend Veblen. He may have revealed some little of his suspicions to Miss Ornstein when he suggested that she take up the matter and determine as well as she might under what auspices the new game was really invented and played. There is no reason, however, to think that she was deflected at any point by the wish to gratify the unworthy *Schadenfreude* of "the professor in charge." All this happened some fifteen years ago, and I hope that many of us see things a little more clearly than in 1912.

Nevertheless, I happen to know of no one who could make any essential improvement in *The Rôle of the Scientific Societies in the Seventeenth Century*. The title is misleading—too diffident, for the book has much to say of the work of individuals, and too restricted since it is properly extended beyond the boundaries set. So far as I know there is no other work which can bring so many edifying reflections into the mind of the scientific worker and at the same time give the intelligent layman so correct a notion of the manner in which our modern science got under way. It is no small achievement to have written such a book on so vital a subject. There is a certain assurance in its march which may have come from the mathematical training to which Miss Ornstein had given preference. There is a lordly quality in mathematics which dismays those of us who are not in the charmed ring. To the historical student there is a fine sense of the pathetic stumbling and groping which has accumulated such "mind" as humanity had thus far achieved and which holds out indefinite hopes for the future.

JAMES HARVEY ROBINSON

AUTHOR'S PREFACE

The history of science may be called a stepchild in the family of the *Natur- und Geschichtswissenschaften*. It is too technical for the historical student, too bookish for the man of the laboratory. Yet if we agree that it is the task of the "new history" to explain what is most vital and fundamental in our civilization today, the historian must incorporate many a chapter of the neglected history of science into his work. For only in this way will he be able to furnish that essential historic background for the achievements of Ehrlich and Mme. Curie that he is wont to give to the projects of Lloyd George and King Ferdinand of Bulgaria. This assimilation and transference of facts from the history of science to general history will naturally fall to those interested equally in the facts of history and the progress of science. As a member of this class, the writer has attempted to describe what seems to her the most vital element in the milieu in which modern science was born.

In books dealing with the histories of the various sciences in the seventeenth century and in treatises touching upon any phase of the intellectual development of the period, a few paragraphs or pages are invariably found emphasizing, on the one hand, that science obtained its most valuable, nay indispensable, aid from the scientific societies of the day; and, on the other hand, that the universities failed to supply such aid. Yet no existing work—so far as the writer knows—tries to show how this aid was given and why the societies were indispensable; nor is there any treatment which follows the work of the universities during this period and points out wherein it was inadequate.

In her attempt to supply in some measure this deficiency the author has encountered serious and to some extent insurmountable difficulties in the lack of *Vorarbeiten* dealing with the university situation, and again in the nature of the subject matter itself, which is so extensive and varied that it has proved well-nigh impossible, within the compass of a dissertation, to do justice to any of the many lines of inquiry. It is thus with a keen sense of its incompleteness that the book is sent to the press.

The author takes this opportunity to express her great thanks and appreciation to Professor James Harvey Robinson who through his writings, teachings, and personal encouragement has been her inspiration throughout this work.

M. O.

March 28, 1913

TABLE OF CONTENTS

PART I

INTRODUCTORY

CHAPTER	PAGE
I. GENERAL SCIENTIFIC ADVANCE IN THE SEVENTEENTH CENTURY	3
II. RÔLE OF INDIVIDUAL SCIENTISTS	21
Gilbert	22
Galilei	23
Torricelli and Pascal	32
Harvey	34
Van Helmont	38
Bacon	39
Descartes	44
Guericke	50
Amateurs	55

PART II

LEARNED SOCIETIES AND JOURNALS

III. ITALIAN SCIENTIFIC SOCIETIES	73
The Accademia dei Lincei	74
The Accademia del Cimento	76
IV. THE ROYAL SOCIETY	91
Origin in London and in Oxford	92
Establishment (1660) and Incorporation (1662)	101
Scientific Apparatus, Methods of Work, and Problems	112
Publication of the Philosophic Transactions	123
Publication of Scientific Works	128
Relation to Newton	132
V. THE ACADÉMIE DES SCIENCES	139
Meetings at Père Mersenne's Cell	142
Establishment by Colbert	144
Method of Work and Problems	148
Decline after Colbert's Death (1683)	156
Reorganization and New Charter (1699)	161

CHAPTER	PAGE
VI. GERMAN SCIENTIFIC SOCIETIES	165
The Societas Ereunetica	167
The Collegium Naturae Curiosum	169
The Collegium Curiosum sive Experimentale	175
Leibniz and the Berlin Academy	177
VII. THE SCIENTIFIC JOURNALS	198
The *Journal des sçavans*	200
The First Medical Journals	202
The *Acta eruditorum*	203
The *Nouvelles de la république des lettres*, etc.	207

PART III

THE LEARNED SOCIETIES AND THE UNIVERSITIES

VIII. SCIENCE IN THE UNIVERSITIES	213
Italy	217
France	220
Germany	226
England	235
Holland	250
CONCLUSION	257
APPENDIX	264
BIBLIOGRAPHY	271
INDEX	287

PART I
INTRODUCTORY

CHAPTER I

GENERAL SCIENTIFIC ADVANCE IN THE SEVENTEENTH CENTURY

To anyone who regards the scientific achievements of our age as the most characteristic and essential elements of modern civilization, and who looks at the present from a historical point of view, the seventeenth century is a period of great significance. For if the progress of a century is shown by a comparison of the knowledge of scientific facts which prevails in its first and in its last decade, no other century, perhaps, can show strides in the realm of knowledge equal to the seventeenth. It not only established the methods and the means of scientific advance; it discovered an immense number of scientific facts; it created a theory of cosmic interrelation to which two centuries have added nothing; it saw, moreover, the commencement of the reaction of newly discovered facts upon the prejudices accumulated through centuries.

Rosenberger in his *History of Physics* says:

Physics before the seventeenth century knew only the methods of natural philosophy and of mathematics. Both had their foundation in the experiences of daily life, in the materials compiled by ordinary observation; but an experimental method which created these independently did not exist. Experiment was used in special instances to measure relations of magnitude of phenomena; an individual inventor might try to win from nature her secrets through experiments, but a systematic questioning of nature, observation as a method in phyics, was not known. The physicist conceived his task as the explanation of known phenomena, but did not see that he had the duty of closer observation, of verification of his hypothesis. The experiment was not part of science; at best it preceded it, but was of no significance in it. Hence wrong statements had little to fear of detection. The realm of thoughts was regarded as infinitely finer than the common material world. Indeed, it would not have been a good sign if philosophic statements fully coincided with experience; and it was no drawback if such statements differed from observation. There was still in natural philosophy something of Platonic revery, of the "idea" and of scorn of matter. The student of natural philosophy thought it beneath his dignity to busy himself, like an artisan, outside of his study and was proud to live in the realm of spirit. Thus it happened that although experiments were made and cleverly made, yet science was little affected by them. It was the task of the seventeenth century to introduce experiment into science and to make the experimental method that recognized in science.[1]

Rosenberger's phrase, "introducing experiment into science," implies indeed many elements. Objectively, it presupposes the production of instruments wherewith to experiment, the creation of places where, and general conditions under which, experimentation can be carried out. Subjectively, it signifies creating in men high standards of exact observation, and developing in them experimental skill. All this the seventeenth century did.

As regards instruments, it produced the microscope, telescope, and machinery for grinding their lenses. It originated an exact time-measuring instrument in the pendulum; it brought into existence the thermometer and barometer, and the air pump. It created therefore the most fundamentally important apparatus of the modern physical laboratory.

The seventeenth century first produced the places where, and conditions under which, experimentation could be carried out. There is one exception to this statement. Chemistry, i.e., alchemy, had had its laboratories, its furnaces, cooling and drying apparatus, mortars, countless glass vessels, distilling contrivances, for many preceding centuries.[2] The apothecary had had his distilling apparatus, his furnace for chemical and pharmaceutical operations.[3] But the conception of a physical laboratory and a non-alchemistic and non-pharmacological laboratory was the creation of the seventeenth century. To be sure the ear-

[1] Dr. Ferd. Rosenberger, *Die Geschichte der Physik,* II, 3 f.

[2] For a picture of an alchemistic laboratory, see *Catalogue of Deutsches Museum* (Munich), p. 270.

[3] The Germanisches Museum at Nürnberg has reconstructed a sixteenth-century apothecary shop in Room 73; see *Guide,* p. 159.

ADVANCE IN THE SEVENTEENTH CENTURY 5

liest laboratories were not very well equipped. The bedroom or kitchen of the scientist was often used as a place for experimentation. Newton's optical researches were made in his lodgings. Robert Boyle tested his laws of elasticity of gases in tubes along the stairs.[4] But before the end of the century such informal workshops of scientists were in some rare instances supplanted by laboratories in the modern sense of the word, supplied with instruments of measurement and with facilities for research work. By 1700 both the chemical and physical laboratory existed in embryonic form.[5]

The astronomical laboratory, the observatory, on account of its affiliation with astrology, existed much earlier. But the seventeenth century created the modern observatory, equipped with telescopes and fine instruments for exact research, prepared for the task of making systematic maps of the celestial regions.[6] The seventeenth century multiplied the establishment of botanical gardens; it insisted upon the erection of anatomical theaters and consequently upon the adoption of methods of dissection in the study of medicine. In most diverse branches of scientific work dwelling places for the cultivation of the spirit of experimentation were established.

The subjective side of Rosenberger's statement, "that the seventeenth century introduced experiment into science," signifies, as has been said, that it produced scientists and scientific skill. The truth of this statement can best be shown perhaps, by comparing, in the various scientific fields in broad outlines, the information of a man familiar with the whole range of science in 1600—whom we shall for convenience call A—with that of a

[4] F. Cajori, *A history of physics in its elementary branches including the evolution of physical laboratories*, p. 288.

[5] The Germanisches Museum at Nürnberg has reconstructed a chemical pharmacological laboratory in Room 76; see *Guide,* p. 163.

[6] The Greeks and Arabians used spheres. Maps of the stars originated in the sixteenth century. For the earliest maps of the stars, see Deutsches Museum, p. 310. For astronomical instruments of the seventeenth century, *ibid.*, p. 316.

man B, in 1700, similarly instructed in the entire scientific knowledge of his time. The difference between the scientific truths in the possession of A and B will then represent, to borrow a phrase from mathematics, the "integration" of the "differential" work and skill of the many individual scientists of the century. Besides, we shall in this way gain a clearer perception of how much the seventeenth century added to the fund of scientific knowledge.

Commencing with physics, and taking up first the fundamental chapter of dynamics, A was permeated with Aristotelian ideas;[7] B, through Galileo, Kepler, and Newton, was in many respects at the level of present-day information. The vast difference this represents may be indicated as follows:

A believed that:

1. Bodies have either a natural motion downward or upward. The former are called "heavy," the latter "positively light."

2. There are two types of motion: that of heavenly bodies is perfect, circular, unchanging; that of earthly bodies is rectilinear and requires for its maintenance a force acting continually. If the force stops, it stops.

3. Bodies fall in accelerated motion because as the body falls the air gives it speed; hence in a vacuum (if conceivable) bodies would fall with uniform velocity.

4. Heavier bodies fall more quickly than light bodies.[8]

B knew that:

1. All bodies are subject to the force of gravitation and are "heavy."

2. Every body, celestial or terrestrial, continues in its state of rest or of uniform motion in a straight line, unless it be compelled by a force to change its state. Uniform rectilinear motion would thus continue forever unless it met resistance. "Force" is that, by means of which rest or motion of a body is changed.

3. Bodies fall in accelerated motion because of the force of gravitation; air does not accelerate but impedes motion.

4. All bodies fall with uniform acceleration.

[7] A knows, however, since Stevin (1585) the laws of motion along the inclined plane.

[8] J. C. Poggendorff, *Geschichte der Physik*, pp. 218 ff.

ADVANCE IN THE SEVENTEENTH CENTURY

Turning from the chapter on dynamics to pneumatics, A could not conceive of the weight of air, or of the creation of a vacuum. "Nature abhors a vacuum" would to him be an axiomatic truth. B would understand the nature of atmospheric pressure (Torricelli); its variation in different weather, at varying altitudes: He would have an air pump and know most of the properties of a vacuum (Guericke and Boyle). In hydrostatics A's knowledge had been started along right lines by Stevin but B would know all fundamental hydrostatic principles (Torricelli, Pascal, Mariotte). In acoustics A would know of the relation of the length of a chord to the pitch; B would understand the laws of vibrating chords, and know the velocity of sound (Mersenne), but still be ignorant of many fundamental truths.

In optics, A would know considerably more than in other fields; for ever since Roger Bacon, the focal properties of spherical mirrors had been understood. Maurylocus (1494–1575) had studied lenses. Then Della Porta's book, *Magiae naturalis*,[9] contained a description of the *camera obscura,* even of a combination of lenses which has been claimed to be the first telescope.[10] B, on the other hand, would be acquainted with the most minute details about the focal properties of lenses (Kepler and Descartes);[11] he would comprehend the laws of refraction of rays passing from thinner into thicker medium (Snellius);[12] he would even be initiated into the phenomenon of diffraction (Grimaldi).[13] He would be aware of the nature of white light and its decomposition into the spectral colors (Newton). He would have learned of the two theories of explaining light: the corpuscular theory of Newton (then accepted), and Huygens' and Hooke's theory of undulation (now accepted).[14]

[9] Johann Baptista Della Porta, *Magiae naturalis,* Book XX (1589).
[10] Poggendorff, *op. cit.,* pp. 129–36.
[11] *Ibid.,* pp. 167–74, 305.
[12] *Ibid.,* p. 311.
[13] *Ibid.,* p. 339.
[14] *Ibid.,* pp. 643, 668, 586 ff.

In magnetism and electricity, A would be acquainted only with the magnet and compass and the electric properties of amber. B, although his knowledge would be much less in this than in the other branches of physics, would nevertheless comprehend the phenomena of terrestrial magnetism, magnetic declination and inclination; he also would be aware of other substances besides amber which exhibit electric properties (Gilbert and Guericke).

In no other science did the seventeenth century, starting from so little, reach so far as in physics; no other science records during the century so many pioneer experimenters.

Turning to astronomy, and comparing the status of the science in 1600 with that in 1700, we realize that astronomy in 1600 had a great start over all other natural sciences. For the Copernican system had been the work of the sixteenth century. But as inherited by the seventeenth century, it contained two fundamental errors: it assumed circular orbits of the planets, and uniform velocity for heavenly bodies. Moreover, in 1600 it was known to few, accepted by fewer; indeed, the *Index* had taken no notice of it as yet. A might thus be an adherent of the Copernican hypothesis, or he might accept Tycho Brahe's compromise cosmic system: that the moon and sun move about the earth, but that Mercury, Venus, Mars, Jupiter, and Saturn about the sun—a system corresponding with observation and having the advantage of not interfering with any biblical passage. Through Tycho Brahe's discoveries A might know of the existence of changeable stars, and how much they outraged those that clung to the perfect unchangeability of heavenly bodies. He might know that Tycho made the unprecedented claim that a comet might be further from the earth than the moon, and that this claim aroused a storm of indignation; for was not the comet's interference in human destinies due to its proximity?[15]

By 1700 the telescope had utterly revolutionized the science

[15] R. Wolf, *Geschichte der Astronomie*, pp. 269 ff.

ADVANCE IN THE SEVENTEENTH CENTURY 9

of astronomy. B, armed with it, could see in the phases of the inner planets definite proofs of the Copernican system. He could perceive the moons of Jupiter and Saturn, and the rings of Saturn. He would know that Kepler, on the basis of Tycho Brahe's observations, had formed empirically his famous laws. Hooke, Wren, and Halley had studied the laws of motion of celestial bodies, and finally Newton proved the truth of the magic formula that attraction varies inversely as the square of distance, thus explaining all astral phenomena. With this Newtonian formula any remnant of rational opposition to Copernicus was removed. By means of it Descartes' elaborate explanation of celestial mechanics as vortex motions was doomed before the century had expired which gave it birth and enthusiastic acceptance. B, in the year 1700, owing to the investigations of Hevelius, Huygens, and Halley, would understand that the paths of comets are subject to the same definite and definable laws as those of planets. He had drifted so far from Aristotelian ideas that sun-spots, varying stars, and the spheroidal shape of the earth, in no way wounded his feeling of the perfection of the universe.[16]

In chemistry A's and B's ideas would not be fundamentally different. A's notion of chemistry in 1600 would be an acceptance of Aristotle's and Paracelsus' ideas. Aristotle taught that there were four fundamental qualities, "humors"—coldness, warmth and dryness (characteristic of the solid), and wetness (connected with the liquid state). These four humors can be arranged in four pairs; and four elements were postulated by Aristotle as the carriers of these pairs of qualities: earth as cold and dry; water as cold and wet; air as wet and hot; fire as dry and hot.[17] In so far, therefore, as these elements were the carriers of

[16] *Ibid.*, pp. 320 ff.
[17] Hermann Kopp, *Bieträge zur Geschichte der Chemie*—Pt. III, "Ansichten über die Aufgabe der Chemie und über die Grundbestandtheile der Körper bei den bedeutendsten Chemikern von Geber bis Stahl," p. 6.

all possible states of matter, they were conceived of as the constituent parts of all the material world. The varying proportion of these Aristotelian elements explained to A the different physical constitution of matter, the different states of aggregation, and changes caused by the influence of temperature. The rapid change of water into ice or into vapor made plausible the assumption that things that seemingly were most dissimilar could yet be of the same substance.

Paracelsus' idea and that of some of his predecessors was that all substances consisted of mercury, sulphur, and salt. This search for and acceptance of three fundamental substances, clung to throughout the seventeenth century, does not seem incomprehensible if properly explained. "Sulphur" does not mean the substance of sulphur, but what is burnable in matter; "mercury," what is volatile, brilliant, metallic; "salt," what remains in the form of ash, after a body is burnt. Every substance was seen to change in fire. It was evidently divided into three parts: that which burns, that which escapes in volatile form, and that which remains as ash. This process was assumed to be necessarily a simplification, a disintegration of the substance into its constituent parts.[18] Hence, by a confusion of thought, the explanation of why all substances thus changed under the influence of fire was found in the assumption that they were composed of these three elements—a typical case of medieval self-deceptive reasoning. Out of these three elements the animal, vegetable, and mineral kingdoms were assumed to be constituted. According to Paracelsus, health was the normal proportion of these three elements, disease the abnormal, which could be cured by medicines so concocted as to rectify the wrong and re-establish the right proportion. With this idea the scope of chemistry was evidently enlarged and emphasis laid upon the importance of making medicines.[19] It became indissolubly connected with the

[18] *Ibid.*, p. 136.
[19] *Ibid.*, p. 135, n. 198, quotes from Paracelsus' book, *Paragranum*, p. 220: "Mache *Arcana* und richte dieselbigen gegen die Krankheiten."

ADVANCE IN THE SEVENTEENTH CENTURY

study of disease and the art of healing.[20] Until the end of the seventeenth century this idea persisted and both A and B might have viewed chemistry's main function as lying in the pharmacist's work, and think in Paracelsian fashion of health and sickness as chemical conditions.

But B might be a follower of Robert Boyle, who stood for a new kind of chemistry, divorced from alchemy and medicine, claiming for it no other function but the investigation of natural phenomena, an end in itself. Boyle insisted that, contrary to Paracelsus' ideas, fire did not reduce bodies into simpler elements; indeed, that varying degrees of heat created different substances, and might even add to substances, while processes other than fire reduced substances into simpler compounds. He further said that only those substances which could not be further divided by any known process should properly be called "elements"; and asserted that there was no definite number of such substances. With this new conception of the elements he laid the cornerstone of modern chemistry. Boyle explained the nature of chemical combination and was the first to see how closely physics and chemistry were in touch in the investigation of natural phenomena. But it must be noted that just as the existence of the Copernican theory in 1600 by no means meant that all astronomers were necessarily Copernicans, so the existence of these advanced ideas of Boyle were hardly accepted by the chemists, and B would normally be a Paracelsian.

As for botany and zoölogy, A would hardly think of them as independent sciences, but as parts of medicine, more especially of pharmacy. Of the two, botany was far in advance. Physicians had by 1600 done extensive botanical work. Great and costly volumes had been published throughout the sixteenth century on

[20] Kopp, *op. cit.*, p. 140: "Er [Paracelsus] leitete hiermit das Zeitalter der medizinischen Chemie ein: die Richtung in welcher die Chemie bis zur zweiten Hälfte des siebenzehnten Jahrhunderts ihre Ausbildung und Repraesentation fand. . . . ."

the description of plants.[21] The discovery of tropical flora in Africa and America had given a great impulse to botanical study and led to the establishment of botanical gardens. Slight attempts had been made at the creation of a classification (Bauhin)[22] and precise nomenclature (Jungius),[23] but the most fundamentally erroneous notions existed. L'Ecluse arranged flowers into two groups: those that smelled sweet and those that had no scent. And Carrichter[24] divided them according to the twelve signs of the zodiac. A good summary of the botanical information of the time may be the works of Cesalpino, professor in Padua, *De plantis*, Book XVI (1583). He still was in what Sachs calls the alchemistic state of botany and spoke of a "soul" residing in the pith of the plant; he thought of sexual organs of flowers only as protecting envelopes, and that "there are some plants [mushrooms] that have no seed .... and spring from decaying substances: they have only to feed themselves and grow and are unable to produce their like."[25]

Zoölogy in 1600 was still in part characterized by the naïve credulity and lack of observation of previous centuries. There existed a branch of "biblical zoölogy," and in 1595 a book of Frey's was printed where animals are described "as they exhort us to virtue and deter us from vice." In 1612 a zoölogy was written for theological students; and in 1675 Athanasius Kircher wrote his *Arca noe*.[26] There could be found in 1600 the same type of encyclopedic compilation of animals as existed of plants (Conrad Gesner), with emphasis laid entirely on a description of external characteristics and with even less attempt at morphological classification than in Aristotle. The dissection of animals was merely an occasional adjunct to medical study.

[21] J. Sachs, *History of Botany, 1530–1860*, p. 5.
[22] *Ibid.*, p. 39.   [23] *Ibid.*, p. 30.
[24] E. H. F. Meyer, *Geschichte der Botanik*, IV, 433.
[25] Sachs, *op. cit.*, p. 54.
[26] J. V. Carus, *Geschichte der Zoologie bis auf Joh. Müller und Charl. Darwin*, p. 309.

ADVANCE IN THE SEVENTEENTH CENTURY

By 1700 the aspect of both sciences had entirely changed. What the telescope was to astronomy, the microscope was to zoölogy and botany. It literally created a new world; it enlivened every drop of water; it led to the discovery of minute organisms, infusoria, and blood corpuscles, and laid the foundation of histology,[27] supplanting apparent uniformity with unlimited complexity. It reversed the former conception that interest in objects was proportionate to their size. B in 1700 would be aware of Grew's and Malpighi's systematic microscopic researches of plants, whereby a new branch of botany—plant anatomy—was founded; of Malpighi's, Leeuwenhoek's, Swammerdam's innumerable microscopic observations of insects and lower forms of animal life. He would know that gradually the study of external characteristics was changing to that of internal structure. John Ray and Willoughby by 1700 had started to create a consistent system of classification of plants and animals based on difference of anatomical structure. With emphasis thus changed, dissection of animals had become the indispensable means of zoölogical study. The truths thus disclosed led Grew to a line of research out of which the science of comparative anatomy was born. In botany a great step in advance must be noted; Grew (1676) and, more clearly, Camerarius (1694) had recognized sex in plants, and the significance of their sex organs, and thus laid the ground for Linnaeus' revolutionary work.

The medical knowledge of A in 1600 rested in the main upon Galen's teachings of physiology. The explanations of the processes of circulation, absorption, and breathing then prevailing were that food absorbed from the alimentary canal is carried by the portal vein to the liver and is "by that great organ" converted into blood endowed with "natural spirits." This blood then goes to the right side of the heart, whence most of it is sent to the body along the veins in a flow followed by an ebb thus securing the nourishment of all organs of the body including the

[27] *Ibid.*, p. 428.

lungs. Some of the blood, however, passes from the right ventricle through innumerable, invisible pores in the septum to the left ventricle where it is mixed with air which is drawn from the lungs as the heart expands. Then by the help of that heat which is innate in the heart, placed there as a source of heat of the body by God, the blood is laden with "vital spirits" and this new kind of blood is again distributed in flow and ebb along the arteries to the various parts of the body, giving them the power of exercising their vital functions. Blood laden with vital spirits reaching the brain generates there a third species of "spirits," the "animal spirits" which—pure and unmixed with blood—are carried along the nerves to bring about movement and carry on the higher function of the body.[28]

By 1700, Harvey's conception of the blood had been promulgated for seventy years. Malpighi's microscopic investigations disclosing the structure of the lungs, the existence of capillary vessels, added the crowning proof to Harvey's teachings,[29] indeed, Malpighi had observed in a frog the actual circulation of the blood.[30] Moreover, within a few years of the publication of Harvey's book, anatomists (Pequet 1651) discovered the existence of the lymphatic system, a deathblow to the Galenic notion that the liver was the place of assimilation of food and blood. Long before 1700 the nature of the process of breathing and its parallelism with burning was understood, and thus another prop of Galen's physiology was overthrown. Besides, the microscope had revealed the structure of the viscera, liver, kidney, and spleen, the structure of muscle, bone, and the existence of blood corpuscles (Leeuwenhoek and Swammerdam). Brain and nerves were studied (Willis) and the sense organs, especially the eye,

[28] Sir M. Foster, *Lectures on the history of physiology during the sixteenth, seventeenth and eighteenth centuries,* pp. 12 f.

[29] *Ibid.,* p. 95.

[30] Published by Malpighi under the caption "I See with My Own Eyes a Certain Great Thing" (*ibid.,* p. 96).

ADVANCE IN THE SEVENTEENTH CENTURY 15

were successfully investigated (Kepler, Descartes). We see therefore that the seventeenth century composed the fundamental chapters of modern physiology. But it tried to do more. Just as in astronomy during the seventeenth century the hypothesis of spirits moving planets had to give way to physical laws of gravitation, so an important school of physicians, the iatrophysicists, looked to physics to explain all life's processes. Borelli and Steno, following Descartes, were the main representatives of this school of thought. By them all movements were explained by laws of mechanics: digestion as purely physical action of the stomach; the exchange of blood in the "capillaries" as simple capillary action; breathing merely as expansion of the bronchial tubes; nervous action as a form of oscillation. In opposition to the iatrophysicists, the iatrochemists, Van Helmont, and De la Boe, following Paracelsus, explained all life as a series of chemical processes; digestion, heart action, breathing, as forms of fermentation. So by 1700 the field of medicine, in spite of its great advances, was rich in controversies, and B was in the midst of a turmoil of conflicting opinions.

In mineralogy A would be in possession of books which remained standard works throughout the seventeenth century.[31] B would know a great deal more in the field of crystallography, owing to investigations of physicists and chemists. For Nicholas Steno and Gulielimini had observed the markings and construction of crystals and the constancy of their angles;[32] Robert Boyle, the crystallization of bismuth;[33] and Bartholinus, the double refraction in Icelandic spar (1670).[34]

In the progress of geology and palaeontology, the seventeenth century is of some importance. A in 1600 could not conceive of sciences which *ex vi termini* were in conflict with the Bible. But he would know of fossils found in rocks, marvelously

[31] George Agricola, *De re metallica*, Book XII.
[32] F. Kobell, *Geschichte der Mineralogie, von 1650–1860*, p. 16.
[33] *Ibid.*, p. 12.   [34] *Ibid.*, p. 7.

like animals, and this resemblance would challenge some explanation. The natural assumption that they were of animal origin was inconceivable; for the church taught that land and sea had been separated on the third day of creation, two days before the appearance of animal life; so rocks could not be crowded with the remains of animals. The explanation offered was that these fossils were "freaks of nature" (*lusus naturae*) having no more connection with living creatures than frost patterns on a window with flowers; or they were styled "figured stones" (*lapides sui generis*), created by some inorganic imitative process within the earth (*spiritus lapidificus*)[35]—*denn wo die Begriffe fehlen*—or at times the explanation of volcanic origin was given to attest the existence of fossils in non-volcanic rocks.

B's information, however deficient, would be somewhat better. By 1700 the peculiar phenomenon presented by fossils and the question of their geological nature had received some attention. The organic origin of fossil forms had to be conceded, and most varied explanations were proffered. Lhwyd (1699) would have them grow from seeds planted from rocks;[36] Hooke (1688) called fossils "manuscripts of nature," and traced their origin to earthquakes;[37] Woodward, who gave all his time to these studies, in his *Essay towards a Natural History of the Earth* (1695), asserted views which defied any reconciliation with the Bible;[38] finally there was Scheuchzer (1699) who tried to explain the difficulty by the fact that during the Deluge animal remains had been deposited—and this theory was sanctioned by the church.[39]

General geological speculation existed in some rare instances. In 1696 a book was published and even reached six editions, which explained the origin of the earth as a comet.[40] Three views showing astonishing insight were evolved during the sev-

[35] K. A. Zittel, *Grundzüge der Palaeontologie (palaeo-zoologie)*, p. 17.
[36] Carus, *op. cit.*, p. 467.
[37] *Ibid.*, p. 23.
[38] *Ibid.*, p. 39.   [39] *Ibid.*, p. 24.   [40] *Ibid.*, p. 40.

ADVANCE IN THE SEVENTEENTH CENTURY 17

enteenth century: Steno (1669) in *De solido intra solidum* asserted that the earth's crust consisted of parallel layers, and that fossils were remnants of organic matter;[41] Leibniz (1693) published a book, *Protogeae*, wherein he described the gradual origin of the spherical shape of the earth, of its waters, its atmosphere, its metals, and its minerals. Here, moreover, he explained along evolutionary lines, the organic origin of fossils—and at the end asserted that this was not in opposition to the Bible.[42] Then there was the above-mentioned essay of Woodward, which contained many views accepted today. But in spite of these isolated instances, a general darkness in matters geological undoubtedly prevailed in B's thoughts.

If in this résumé of scientific achievements of the seventeenth century we include mathematics, it is not from the point of view of mathematics as an independent discipline. For this stands closer to pure philosophy than to experimental science. Only so far as it belongs in the category of instruments of scientific research will it be noticed here, and we would ask in what way it differed, when wielded by A, from the tool in the hand of B.

By 1600 new computation in the fundamental processes of arithmetic had been reduced to fairly manageable form; calculation with fractions had become simplified, and even the decimal point had been introduced. As regards geometry A had not passed fundamentally beyond the wide knowledge of the Greeks, typified by Euclid's elements and Apollonius' conic sections. He understood the trigonometric functions, and had for his astronomical calculations trigonometric tables carefully computed—a valuable legacy of the sixteenth century.[43] In algebra he had all the fundamental notions, though not as yet expressed in modern terminology.

[41] *Ibid.*, p. 36.
[42] K. Fischer, *Gottfried W. Leibniz*, p. 188.
[43] Dr. Karl Fink, *A Brief History of Mathematics*, pp. 39 f.

B's knowledge comprised only an elaboration and simplification of facts previously known. In the field of arithmetic, the seventeenth century, characteristically, first ventured upon the invention of calculating machines.[44] It originated the study of convergent series, thence evolved logarithms (Napier, Briggs), and thereby produced a revolution in the field of computation.[45] B would know of new practical applications of arithmetic for statistical purposes, such as mortality tables (Sir W. Petty, Halley);[46] of the theory of probability (Pascal, Fermat);[47] and would have his algebra in modern form (Descartes). He would be acquainted with Descartes' method of expressing geometric conceptions by algebraic equations, referring points to a fixed system of co-ordinate axes. This invention proved an invaluable help to exact thought and expression, a photographic method, so to speak, of picturing interrelations of magnitudes such as frequently occur in physics, chemistry, and indeed in every realm of the physical world. B would be initiated in the study of the higher plane curves (Descartes, Leibniz, Bernouilli, Newton); he would be aware of the many properties of cycloid curves (Galileo, Roberval, Huygens, Bernouilli), a curve of the greatest importance in the history of science, as though its properties Huygens discovered his isochronous pendulum.[48] But the researches which were destined to make mathematics henceforth the most powerful tool in science dealt with the use of the infinitesimals. At first this was studied for the sole purpose of calculating volumes and areas (Kepler, Cavalieri, Roberval, Huygens, Wren, Wallis, Barrow); later, in the hands of Newton and Leibniz, it became the differential and integral calculus.[49] Thereby mathematics was fashioned into the supreme instrument of research in theoretical physics and astronomy. In this form, it bears to their study the same relative importance as the telescope to astronomy, and the microscope to zoölogy, even more

[44] Ibid., p. 48.
[45] Ibid., pp. 288 ff.
[46] Ibid., p. 57.
[47] Ibid., p. 148.
[48] Ibid., pp. 228–40 passim.
[49] Ibid., pp. 168 ff.

ADVANCE IN THE SEVENTEENTH CENTURY

perfect in the absolute exactness of its responses to the scientist's questions. This instrument of scientific research was destined well nigh to monopolize the efforts of scientists of the eighteenth century, and, allied with methods of direct experimentation, held and is still holding sway in the study of the inorganic sciences.

To sum up: It has been pointed out that Rosenberger's phrase, "introducing experiment into science," implies, when taken subjectively, creating in men a high standard of exact observation and developing in them experimental skill. That the seventeenth century did create and develop this I have attempted to show, not by pointing to a long line of scientists who lived then, but rather by comparing the information of two imaginary characters assumed to be acquainted with the entire range of the science of their day. The comparison has made clear that great progress—though in varying degrees—was made in all the different sciences. The greatest progress was evinced in physics, astronomy, medicine, and mathematics, in which indeed the fundamental facts upon which these sciences have been further developed were established; considerable progress was shown in botany, zoölogy, and chemistry; least in geology and paleontology. All this scientific advance, with the exception of that in mathematics, was won by experimentation and observation. Every new truth was explained on the basis of demonstrable facts before it came to be incorporated in the body of scientific knowledge. Indeed, this vast progress in science represents, as even this superficial survey has shown, the summation of individual efforts of many men, and proves conclusively the "subjective" side of Rosenberger's contention, that the seventeenth century developed ability and high standards of observation and experimental skill.

Thus the seventeenth century stands out as the century that introduced experiment and thereby dynamic changes into science. This was in striking contrast with other phases of mental activity of the time. There was little evidence of a general clearing-away of old superstitions. Belief in witchcraft was almost universal, and in the last decade professors at the most enlightened university of Germany—Halle—met to discuss the question of witch trials, and Christian Thomasius, one of the leaders of thought, joined the debate, fully convinced of their necessity.[50] The belief in the efficacy of the "touch" held sway into the eighteenth century; Louis XIV touched sixteen hundred persons, Charles II even more.[51] Boyle, the experimenter, traveled to Ireland to be "touched" by Valentine Greatrix.[52] Just as firmly rooted was the belief in sympathetic and magnetic cures and powders. The literal acceptance of the "Christian epic" was adhered to by the greatest scientists, Boyle and Newton, and a great deal of the mental energy of the century was consumed in explaining, on the one hand, the identity of Christian tradition and the new truths obtained through experimentation; and, on the other hand, in asserting the "innocence" of these studies and showing that they did not interfere with the orthodoxy of religious creed. By a strange "division of labor" it remained for a non-scientific set of thinkers, the deists, to let the scientists' discoveries react upon the Christian epic, and to commence the subversion of the Bible as the court of final appeal in scientific matters.

[50] Andrew White, *Seven great statesmen in the warfare of humanity with unreason*, p. 138.
[51] Charles R. Weld, *A history of the Royal Society with Memoirs of the Presidents, compiled from Authentic Documents*, I, 89.
[52] *Ibid.*, p. 90.

CHAPTER II
RÔLE OF INDIVIDUAL SCIENTISTS

We have heretofore spoken of the seventeenth century as a unit; but a dividing line may be drawn at about the middle of the century, and a closer analysis will reveal that the forces at work during the latter half of the century were different from those which produced the scientific achievements of the first half. This first half seems more like a "mutation" than a normal, gradual evolution from previous times. It accomplished through the work of a few men a revolution in the established habits of thought and inquiry, compared to which most revolutions registered in history seem insignificant. It created the experimental method, it invented and used with startling results the telescope and microscope; it exhibited the vanity and insufficiency of a great part of the traditional knowledge. The second half of the century elaborated these results. Much of this elaboration was accomplished by science-loving amateurs, who often in enthusiastic co-operation co-ordinated their efforts, devoted themselves to experimentation, and to the creation and improvement of instruments, who indeed—to use Rosenberger's phrase again— "reduced science to a worship and idolization of experiment as an end in itself."[1]

A complete analysis of all the forces which created the change from 1600 to 1650 would be too far reaching. We should have to give an account of the reasons why preceding centuries were in the main satisfied to hand down uncritically the scientific heritage of the Greeks, unchanged but for the augmentations and elucidations of the Arabian thinkers; and further still, we should have to explain why the Greeks left their heritage in a form

[1] Rosenberger, *op. cit.*, II, 135.

which did not have "the seminal living principle" within itself. No such complete analysis will be undertaken here; but merely in cursory fashion the attempt will be made to review the work of the two types of men who contributed to this change: (*a*) the scientists, who did the pioneer work of showing the insufficiency of the facts handed down from the past, who established experiment definitely as the chief means of successful scientific progress, who invented the telescope and microscope and used them to prove positively that the store of inherited knowledge was capable of indefinite extension; and (*b*) the philosophers, the propagandists of this movement, who revealed the illusiveness of the methods of study hitherto followed, and who preached as a new gospel those very modes of inquiry which the scientists had adopted.

First in time among the scientists who were to usher in a new era was Dr. William Gilbert (1540–1603),[2] the learned physician of Queen Elizabeth. His work, *De magnete magneticisque corporibus et de magno magnete tellure physiologica nova* (London, 1600), was the first book that contained nothing of peripatetic natural philosophy, did not despise observation in deference to authority, but which was based entirely upon experiment, and showed great skill in the use of the experimental method in the investigation of new phenomena. Gilbert's observations of the magnetic needle led him to the conclusion that the earth was a vast magnet. In order to prove this, he constructed a large spherical magnet, which he called *terella* ("little earth"), suspended a magnetic needle near it, and found the closest analogy of its action to that of a magnetic needle near the earth—indeed, he determined poles, meridians, and the Equator upon the *terella*. By this experiment he proved his conjecture, employing the typical methods of a scientist: first, hypothesis; then construction of apparatus to prove the hypothesis; and then proof

[2] F. Dannemann, *Die Naturwissenschaften in ihrer Entwicklung und in ihrem Zusammenhange*, II (1911), 85–92.

by experiment. From his work on magnetism, Gilbert passed to the observation of what are called today "electric phenomena." Up to Gilbert's time it was known that amber attracted light bodies. He first sought to determine whether other bodies exhibited similar powers of attraction, and found a long series of objects showing the same phenomena. Considering this attraction and independent force in nature, he called it "electricity," and thus became the father of this branch of physics. Furthermore he speculated in an interesting way on the differences between magnetism and electricity.

In his other work, *De mundo nostro sublunari philosophia nova,* Gilbert showed himself in direct opposition to the philosophy of Aristotle. He rejected the notion of "levity" or positive lightness, accepted the Copernican system, and "seeing everything under magnetic aspect," as Francis Bacon asserts, he postulated a magnetic force interacting between the stellar bodies.

In spite of Dr. William Gilbert's brilliant hypotheses and experiments, the entire accomplishments of these decades seem concentrated in Galileo.[3] He belongs among that small group of men such as Petrarch, Erasmus, and Voltaire, whose lives stand for a transitional epoch in the history of men's mind, who at once bear the marks of an age they help to supplant, and in their thoughts create and anticipate the development of succeeding ages. Galileo's popular fame is based on his work in astronomy and his suffering for the Copernican doctrine. But his reputation as the first modern scientist is based on his contribution to the overthrow of Aristotelian physics, his introduction of new methods of investigation, and his development and use of those instruments which did the most to "advertise" the cause of science. Nurtured in the principles of Aristotelian mechanics, which had been accepted for nineteen hundred years, he was led, by observation, to doubt them. Having dared to doubt, he soon disproved

[3] For good account, see *ibid.,* pp. 15–70.

them in a fashion typical of the man who stood at the juncture of two opposing schools of thought; he disproved them by scholastic reasoning and by experiment. Let us follow him in two characteristic instances of his famous pioneer work in mechanics.

Galileo was of course taught that heavy bodies fall faster than light ones, but he saw all the candelabra in the dome of Pisa, heavy and light, swing in equal periods, and as he conceived of this swinging of the pendulum as a form of falling (the inference of a genius) he saw plainly that Aristotle's idea of the relation of weight of body and time of fall was at variance with observable facts, and thus came to enunciate the law that all bodies fall with equal velocity. Then his mind, in medieval fashion, evolved two speculative scholastic proofs for his surmise and the refutation of Aristotle's principles.[4] With these speculations the medieval mind would have stopped. But because Galileo sought experimental proofs for his truths, he differed from his contemporaries. He made two hundred trials, dropping heavy and light weights from the tower of Pisa, and, where resistance of air did not cause differences, he found they reached the ground simultaneously. Later, in Padua, he studied experimentally pendula of different weights, and found them vibrating in the same periods. Then, and only then, was he convinced that Aristotle's law of free fall

[4] Galileo Galilei, *Discorsi e dimostrazioni matematiche intorno a due nuove scienze*. One of the interlocutors says to the other: "Conceive of a falling mass, disintegrated into a number of small particles. All these particles, being of equal size, will reach the earth simultaneously; yet conceived as a unit they are the heavy mass, and this demonstrates that the entire mass falls at the same rate as its component light particles, hence the time of fall is independent of weight. Again, suppose a heavy and a light body dropped simultaneously. The time occupied by their fall must be, according to Aristotle, a mean between the shorter time occupied by the fall of the heavy and the longer time by the fall of the lighter body. But as their combined weight is greater than that of either, the time occupied in the fall of both ought to be less than that used in the fall of the heavy body. The incompatibility of these results showed Aristotle's conclusions to be wrong."

was erroneous, and that his own assumption that all bodies fall in equal time was proved.

Turning to another experiment of Galileo: Aristotelian physics had recognized the acceleration of motion of falling bodies. Galileo sought for a mathematical formula of this acceleration. He started with the arbitrary medieval assumption that this acceleration would be most "regular" and "normal" if it were uniform.[5] But Galileo recognized that this was a pure surmise, and that its correctness depended upon his success in its experimental demonstration. Here difficulties arose at once. "Velocities" could not be measured. With masterly analysis he deducted from the assumed law $v:v' = t:t'$; the other that $s:s' = t:t'^2$,[6] which proportion had the advantage that its terms were measurable quantities. But still the time occupied in falling was under normal circumstances too short to be measured with accuracy by instruments at Galileo's disposal; therefore, by a feat of scientific imagination he overcame this seemingly insurmountable difficulty. By a process of reasoning, which cannot stand today's scrutiny, though the conclusions are correct, Galileo concluded that a ball rolling along an inclined plane was subject to the same laws as a body falling from the height of the inclined plane; only, that by making the hypotenuse of such an inclined plane, say, twelve times longer than the height, the duration of time occupied by the "fall" was increased twelvefold. With such an inclined plane he proceeded to study the laws of free fall. These experiments and the apparatus employed are of particular interest in the history of science.[7] A sphere of brass, highly polished, rolled along the inclined plane in a ridge bordered with parchment, on which there was a scale. To measure

[5] I.e., if the velocity in each successive second would receive equal increments and so be proportional to time.

[6] Where s stands for distance and t for time.

[7] For models of the instruments, see Deutsches Museum (Munich), p. 212.

the time occupied in the "fall" Galileo attached a very small spout to the bottom of a water pail; the water escaping through the spout, during the time when the body traveled through a given distance, was collected in a cup, and its weight served as the measure of time. With this apparatus Galileo found that his hypothesis—that the velocity in successive seconds received equal increments—was correct. The fundamental laws of free fall, of bodies acted on by a constant force, of uniform acceleration, etc., were established.

I have singled out these two instances to show Galileo not only as a theorizer in dynamics, but as "the experimenter," and the originator of the entire scientific method. In histories of the microscope[8] and telescope,[9] conjectures can be found as to which particular grinder of lenses in Holland first conceived of these instruments. But it is generally admitted that it was Galileo's microscope which was first used by Stelluti of the Accademia dei Lincei for systematic research;[10] and it was Galileo's telescope which was first directed to the heavens.

It was almost an accident that led Galileo to the discovery of the telescope.[11] With this instrument, which magnified at first three, then eight, and finally thirty-three diameters, he made the first telescopic astronomical observations. These he published in the *Sidereus nuntius,* one of the most famous short pamphlets in the history of experimental science. The enthusiastic author reveled in the novelty of what he had found, in the wonder of his new *organum* ("instrument"), and in the verification of the Copernican theory. In his dedication to Cosimo II he says: "I

[8] Petri, *Das Mikroskop, von seinen Anfängen bis zur jetzigen Vervollkommung.*

[9] Servus, H. *Geschichte des Fernrohrs bis auf die neueste Zeit.*

[10] See below.

[11] While in Venice he heard a report of the invention of a telescope, and surmising its construction by his experimental genius he promised the authorities to fashion one. He succeeded, and received from the appreciative senate a life-pension.

shall in this small tract bring great news to those who are in the habit of contemplating nature. Things great not only on account of their beauty, but also on account of the instrument through which they presented themselves." He told of his new observations of mountains on the moon, of the four moons of Jupiter, "the Medicean Planets" and their varying phases. He exulted in the proof thus supplied of the Copernican system ("because they revolve about Jupiter just as Venus and Mercury revolve about the sun"). In addition, the old argument that all stellar bodies must revolve about the earth because the moon did, lost its force, inasmuch as the "Medicean" planets revolved about Jupiter. The little book *Sidereus nuntius* did more than bring new light upon astronomical matters and support the Copernican doctrine; its effect went far beyond the province of astronomy. It tended to dissociate in the minds of its many readers the experimental study of nature from the prejudices popularly held against anything that seemed connected with the dark methods of alchemists; it demonstrated that this new method of inquiry, this scrutinizing of nature by new contrivances, might reveal to man deep truths hitherto unsuspected.

Galileo made a tour of Italy with the telescope, to show his new discoveries. Cosimo II appointed him court mathematician, and gave him the professorship of mathematics at Pisa without the obligation to teach. There Galileo continued his astronomical work, observed the phases of Venus (final proof of the Copernican system), and sun-spots. This latter discovery proved unfortunate, because it engaged him in a priority contest with the Jesuit Scheiner, and excited the hostility of the order against him. At the same time the Dominicans began to attack the Copernican system, in the seventieth year of its existence. Galileo as its defender found himself more and more involved. In a letter to the mother of the Duke of Tuscany which has become famous, Galileo refutes most eloquently the charge that the Copernican doctrine is heretical:

We bring new discoveries not to confuse the minds, but to enlighten them, not to destroy science, but to put it on a sound foundation. Our opponents call that false and heretical which they cannot refute, and use their feigned religious zeal as a shield and the Bible as a servant of their private designs. He who would cling to the literal interpretation of the Bible would find contradictions in such expressions as the "eyes, hand, or wrath of God." But all this is adapted to the conception of the people. . . . . The Bible speaks as the people of the time looked upon matters. . . . . In science man must begin not with the authority of the Bible, but with observation and proof. . . . . The Bible cannot be at variance with facts because God cannot contradict Himself. It were risking the authority of the Bible, if when once the facts are proved, the Bible were not interpreted to fit these facts, rather than that man should go counter to the facts and proofs of nature.[12]

In 1616 Galileo[13] went to Rome to justify his teachings. But in spite of a friendly interview with Pope Paul V, the Congregation of the *Index* first condemned the Copernican system, February 25, and later, on March 5, Galileo's teachings.[14] Galileo returned to Florence and lived there, engaged in his researches until he wrote his *Dialogo sopra i due massimi sistemi del mondo Tolemaico e Copernicano* (1632), the work which, through its condemnation by the inquisition, ended his career as an astronomer. It was written in Italian, and contained a debate between the old and new science, between the physicist armed with the text of Aristotle and the experimenter armed with the telescope.

The three characters[15] of this famous dialogue are masterfully drawn. Salviati,[16] at one time a merchant and senator of Florence, a pupil of Galileo—a learned man, member of the Academia dei Lincei—stands for the progressive scientist and voices Galileo's convictions; Nicolo Sagredo,[17] a man much interested

---

[12] Dannemann, *op. cit.*, II, 22, gives the German translation of the letter quoted from Carrière, *Die philosophische Weltanschauung der Reformationszeit.*

[13] Emil Strauss, *Dialog über die beiden hauptsächlichen Weltsysteme, das Ptolomäische und Kopernikanische von Galileo Galilei* (a German translation of the *Dialogo*), pp. xxxv ff.

[14] This is, however, debated.   [16] Salviati died in 1614.
[15] Strauss, *op. cit.*, pp. xlix ff.   [17] Sagredo died in 1620.

RÔLE OF INDIVIDUAL SCIENTISTS 29

in science and mathematics, though in an amateurish way, represents the educated layman, favorably disposed to progress. The third character is Simplicio, the representative of conservative, authoritative book knowledge, a figure suggesting Wagner in Goethe's *Faust*. He tries to understand his opponents and wants to learn of the new teachings, not in the least afraid that he may be converted, certain that he will be only the more fortified in his views and the better able to refute and strangle these newfangled and dangerous doctrines.

As we open the book, we see the frontispiece characteristically representing three men in argument, Aristotle, Ptolemy, and Copernicus. In glancing at the first pages of the dialogue, the timidity and insincerity evidently demanded by existing circumstances are at once keenly felt. In the dedication the author emphasizes the fact that Copernicus' view had been condemned, but that nevertheless he wishes to show the reader that in Italy and Rome as much is known about Copernicus as anywhere. Salviati, a most unreserved defender of Copernicus' views, says he is of course only playing the Copernican, using as it were that mask in the comedy. As the dialogue proceeds, it does not take the form of an exposition and defense of the Copernican theory, but is a refutation of arguments brought by Aristotelians, amateurs, and scientists against Copernicus. Incidentally it purposes to popularize the scientific point of view and give the reader a notion of the true method of "modern" science. It strikes at the heart of the Schoolmen's deference for Aristotle.[18] The Aristo-

---

[18] As for instance when Simplicio asks, "If Aristotle is not the guide, who then should be? Name the authority." Salviati replies: "We need a guide in unknown countries; in known places only the blind need direction; and the blind would better stay at home. He who has eyes, better take them as guides" (Strauss, *op. cit.*, p. 117). Again Salviati says: "It is disgraceful at a public disputation, where things which can be proved are being dealt with, to have someone bring up a quotation of Aristotle. Bring us, Simplicio, your own or Aristotle's proofs; for we are dealing with the world of senses and not of paper" (*ibid.*, p. 118). In another passage Salviati says: "If the subject of our discussion were a question of jurisprudence or human affairs, eloquence would be in place. In sci-

telian idea of the fundamental difference of heavenly and earthly mechanics is shown to be worthless. He points out that there is no difference between celestial and terrestrial bodies; that all move in circles, that all are equally changeable, as sun-spots and variable stars show; that the heavenly bodies are not "perfect" spheres, as may be seen by the existence of mountains on the moon.

The purely physical and astronomical arguments for and against the Copernican system are then reviewed. The Medicean stars, whose periods are forty-two hours, three and a half days, seven days, sixteen days, according to their nearness to Jupiter, are given as illustrations of the motions of other members of our solar system, the nearest planets having the shortest periods of rotation.[19] Interspersed with physics and mechanics, we find a keen psychological analysis of the forces which seemed to Galileo responsible for mental stagnation. He bemoans man's impudence in taking his conceptions as the measure of all things. Salviati compares Simplicio, who would know the purpose of a

---

ence, where conclusions are true and necessary, where nothing is arbitrary, words will not help. Wise men can't spite nature. Sun-spots are facts" (*ibid.*, p. 57). Salviati urged that no one who followed the process of dissecting a body would hesitate to admit that the nerves ran to the brain, if this idea were not rejected by Aristotle. One of Aristotle's admirers had gone so far as to claim that it was the learned Greek himself who invented the telescope. "People injure Aristotle," Salviati urged, "by too much belief; he would have adopted the modern point of view had he had a telescope" (*ibid.*, pp. 113 ff.).

[19] Much space is given to the following argument, conceived as the strongest point against the rotation of the earth. It was claimed that if a body were thrown from the mast of a moving ship, it would not fall at the foot of the mast, but some feet away, owing to the motion of the ship. Hence, it was argued, that a ball dropped from the Tower of Pisa would, if the earth rotated, not strike at the foot, but far away. Salviati proves that this assertion about the ship is a pure assumption; that, in as much as the object dropped partakes of the motion of the ship, the fall would be (and is) as if the ship were at rest. Similarly the earth can rotate and yet the stone fall at the foot of the tower (Strauss, *op. cit.*, p. 145). Incidentally he adds: "We have as little conception of what makes a stone fall as what keeps the moon in its course" (*ibid.*, p. 249).

system like the Copernican before accepting it, to the berry that thinks the sun is created only to ripen it.[20] He points out that the fact that the Copernican theory is opposed to our primitive sense perceptions, our intuitive surmises, should not condemn it. Man is far too ready to accept preconceived notions, and shape his premises to fit his conclusions.[21]

There is nice psychological analysis in a comparison drawn between the adherents and opponents of the Copernican theory. The Copernican's belief is the result of argument and study: "He who gives up received opinions in which he was nurtured, for those accepted by but few persons, rejected in the universities, seemingly paradoxical, must be influenced—nay, forced, by good reasons, while their opponents need not have studied the matter at all." Simplicio confesses that he has not as yet read many of the new books; he hasn't tested the telescope, but in the face of the opinion of men who have tested it a hundred times, he asserts he hasn't much confidence in it, and that many things seen through it may be but illusions created by the lenses.[22]

No part of the *Dialogo* is more interesting than that which deals with Gilbert's discovery. Galileo seemed conscious that in him he had a *Mitstreiter* not only in the Copernican controversy, but in support of the entire modern scientific attitude. The magnet's threefold motion is, to Galileo, proof positive that the earth can have such a threefold motion—for the objection of the Aristotelians that the earth is simple and the magnet compound seems to Galileo as if one said bread was simple but one of its ingredients compound.[23]

That the *Dialogo* was written as a means of propaganda for the new science seems evident from the fact that it was written in Italian. It was this fact in great part that raised a storm of protest which no other scientific work except Darwin's *Origin of Species* has ever raised. The Pope, told that he was caricatured

[20] *Ibid.*, p. 384.
[21] *Ibid.*, p. 279.
[22] *Ibid.*, p. 369.
[23] *Ibid.*, pp. 430 ff.

in the figure of Simplicio, was merciless, and the ensuing trial and condemnation form the most thrilling chapter in the *Warfare of Science and Theology*. The condemnation had far-reaching results. It could not stop scientific inquiry, but by committing the clergy and the Catholic universities to the side opposed to experimental science, it clearly defined the issue in the "warfare" between scientific truth, on the one hand, and church and Bible, on the other—the last scenes of which we are still witnessing. Further, it implied that science must seek its support outside of the university, a point of great importance for the future both of science and of the universities.

As with most figures in periods of transition of thought, Galileo was strangely conservative on a few points. For instance, he accepted in Aristotelian fashion the *resistenza del vacuo*, a modified *horror vacui*, as an explanation of why a pump could raise water only thirty-two feet.[24] In the *Discorsi e demostrazione matematiche* Galileo says that, as in the case of a suspended coil of wire there is a length at which its own weight breaks it, so it must be with the column of water raised by the pump.[25] Inasmuch as Galileo knew that air had weight, and had devised a means of weighing it, all this is the more strange,[26] and in a measure enhances the historical interest of the man.

It was appropriate that Galileo's pupil and successor as court mathematician of the Medici, Torricelli (1608–47), should discover the principles underlying the *horror vacui*, and in this discovery may be seen a triumph of the experimental method. Torricelli,[27] finding the water column too high to manipulate,

[24] The story goes that when a Florentine gardener informed him that he could raise water, by means of a pump, only to a fixed height, and not beyond it, Galileo, at first amazed, declared that the *horror vacui* had apparently this definite limitation.

[25] *Discorsi* in Ostwald's *Klassiker der exacten Wissenschaften*, No. 11, p. 17.

[26] Dannemann, *op. cit.*, p. 39.

[27] *Ibid.*, pp. 159–61.

RÔLE OF INDIVIDUAL SCIENTISTS 33

decided to determine the height to which he could raise mercury, and noticed it to be twenty-eight inches (a column having of course the same weight as thirty-two feet of water). This in itself seemed to indicate nothing but that the same degree of *horror vacui* acted upon mercury as upon the water in the pump. But the fact that this careful observer noted that the height of his column varied on different days gave a clue to its connection with the atmosphere, for "nature," as he put it, "would not, as a flirtatious girl, have a different *horror vacui* different days." Strangely, he did not interest himself so much in the question of the vacuum itself as in that of the varying height of the column of mercury, and in the construction of the barometer. Pascal (1623-62), on hearing of Torricelli's experiments, as final proof of the connection between the phenomenon and atmospheric pressure undertook to determine whether the height of a column of mercury would be different on top of a mountain than at its foot. The interesting letter in which he urged his brother-in-law, Périer, to make the experiment on the Mount Puy de Dome (970 m.) has come down to us with Périer's answer, giving a detailed description of how he measured the mercury column in the valley, and how he repeated the measurement on top of the mountain—finding, to his surprise, a very great difference. Torricelli, however, making similar experiments at the duomo of Florence, claimed priority in this train of thought.[28] In addition, Torricelli and Pascal discovered the laws of hydrostatics, and both must be counted among the great experimenters of the time.

Kepler[29] (1571-1630), who stands next to Gilbert and Galileo as a pioneer in physical science, differs from them, when his work is considered as a whole, in being rather a mathematical than an experimental physicist. But in his work in optics and with the telescope, he belongs with them; and as the telescope is

---

[28] Blaise Pascal, *Récit de la grande expérience de l'équilibre des liqueurs* (Paris, 1648).

[29] Dannemann, *op. cit.*, pp. 114-33.

the instrument which "advertised" the new science most, he must ever be ranked among the pioneers that marshaled the forces of the new science. The discovery of the laws which bear his name links him inseparably to those branches of mathematics by which in the eighteenth century physics was destined to conquer the skies. His work, *Dioptrice* (1611), though its conclusions were based only upon most primitive methods of experimentation, was the basis of the great optical experiments of the seventeenth century.

Turning to the pioneers of experimentation in organic science, i.e., medicine, during the years 1600–1650, the name of Harvey stands out as Galileo's does in physics. But there is one important point of difference: Galileo, as we have seen, was the first to perceive that Aristotle's mechanics were wrong, and to substitute new laws. In medical science a line of predecessors of Harvey had done the pioneer work of refuting Galen's omniscience. Vesalius[30] (1515–64) had first insisted on methods of dissection, and had at first timidly, then more and more boldly, proclaimed that what he saw at the dissecting table did not always tally with Galen's teaching, and that Galen was wrong. He incorporated his results in a book, "Structure of the Human Body" (1543),[31] and thereby became the founder of the non-Galenic science of anatomy. Vesalius' work was continued by his pupils, Falloppio (1523–62) and Realdus Columbus (1516–59). Ceasalpinus (1519–1603), a pupil of Falloppio, understood the pulmonary circulation of the blood, and Fabricius (1537–1619), the successor of Falloppio as professor of anatomy in Padua, wrote a book with correct views on the valves of the heart.[32] Harvey (1578–1657), Fabricius' pupil, welded together the several links which these men had furnished and was indeed—according to Foster—not the first to discover, but to demonstrate, the cir-

[30] For a full account of Vesalius, see Foster, *op. cit.*, pp. 1–24.
[31] *De humani corporis fabrica.*
[32] Foster, *op. cit.*, pp. 25–41.

RÔLE OF INDIVIDUAL SCIENTISTS 35

culation of the blood. His immortal work, *Exercitatio anatomica de motu cordis et sanguinis in animalibus*, is one of the most interesting books of this period to the historical student, not only on account of the intrinsic importance of the discovery it contains, but because it frankly tells of the train of thought which led to the discovery and admits the reader, as it were, to the mental workshop of the author. Harvey was conscious of the difficulties of his problem, and of the fact that only experimental methods would help to solve it. He wrote as follows:

> When I first gave my mind to vivisection as a means of discovering the motions and uses of the heart, and sought to discover these from actual inspection, and not from the writings of others, I found the task so truly arduous, so full of difficulties that I was almost tempted to think . . . . that the motion of the heart was only to be comprehended by God. . . . . At length, and by using greater and daily diligence and investigation . . . . making frequent inspections of many and various animals, and collating numerous observations, I thought that I had attained the truth . . . . and that I had discovered . . . . both the motion and use of the heart and arteries.[33]

He first investigated the true nature and purpose of the movements of the heart itself and found "that the motion of the heart consists in a certain universal tension, both contraction in the line of its fibres and constriction in every sense. . . ." and that "the heart when it contracts is emptied. Whence the motion which is generally regarded as the diastole of the heart is in truth its systole."[34] The pressure of the constriction squeezes the blood into the arteries, and it is this transmitted pressure which causes the pulse.[35]

With this changed point of view, through observation of

[33] William Harvey, *On the Motion of the Heart and Blood in Animals* (Willis' trans., revised by Alex. Bowie), chap. i, p. 20.

[34] *Ibid.*, pp. 23 f.

[35] *Ibid.*, p. 26. The arteries do not swell in order to suck in the blood but because blood is driven into them by the systole of the heart. "They are filled like sacs . . . . and do not expand like bellows."

man, experiments on hearts of fishes, frogs, birds, and embryos,[36] Harvey was led to a correct conception of the functions of the auricles, ventricles, and their respective valves,[37] dispensing definitely with Galen's "invisible pores" in the septum of the heart. This in turn led him to a true understanding of the pulmonary circulation of the blood.[38] Then applying to the greater circulation the same conclusions as those at which he had arrived in regard to the pulmonary circulation, he arrived at a view ".... of a character so novel and unheard-of that I [Harvey] not only fear injury to myself from the envy of a few, but I tremble lest I have mankind at large my enemies."[39] This novel view was of course the circulation of the blood. It is very interesting to note that Harvey was led to his views on the circulation of the blood, not by speculative deductions, but by the consideration of the quantitative features of blood. He writes:

I frequently and seriously bethought me, ... what might be the quantity of blood which was transmitted, in how short a time its passage

[36] *Ibid.*, pp. 29 ff.; chap. vi, p. 37: "Dissections of animals must furnish the data necessary to solve the problem. .... They plainly do amiss who, pretending to speak of the parts of animals generally as anatomists for the most part do, confine their researches to the human body alone, and that when it is dead."

[37] *Ibid.*, p. 32: ".... the auricle contracts, and forces the blood .... into the ventricle .... which being filled, the heart raises itself straightway, makes all its fibres tense, contracts the ventricles, and performs a beat by which beat is immediately sent the blood supplied to it by the auricle into the arteries. The right ventricle sends its charge into the lungs by the vessel which is called *vena arteriosa*, but which in structure and function and all other respects, is an artery. The left ventricle sends its charge into the *aorta*, and through this by the arteries to the body at large."

[38] *Ibid.*, p. 42: "In the warmer adult animals, and man, the blood passes from the right ventricle of the heart by the pulmonary artery, into the lungs, and thence by the pulmonary veins into the left auricle, and from there into the left ventricle of the heart."

[39] *Ibid.*, chap. vii, p. 47: .... "so much doth wont and custom become a second nature. Doctrine once sown strikes deep its root, and respect for antiquity influences all men. Still the die is cast, and my trust is in my love of truth and the candor of cultivated minds."

RÔLE OF INDIVIDUAL SCIENTISTS

might be effected, and the like. . . . . But not finding it possible[40] that this could be supplied by the juices of the ingested aliment without the veins on the one hand becoming drained, and the arteries on the other hand becoming ruptured through the excessive charge of blood, unless the blood should somehow find its way from the arteries into the veins, and so return to the right side of the heart, I began to think whether there might not be a motion, as it were, in a circle. Now this I afterwards found to be true and I finally saw that the blood, forced by the action of the left ventricle into the arteries was distributed to the body at large, and its several parts, *in the same manner as it is sent through the lungs,* impelled by the right ventricle into the pulmonary artery, and that it then passed through the veins and along the vena cava and so round to the left ventricle. . . . .[41]

From this reasoning and numerous dissections of animals and embryos,[42] and observation of pathological conditions, all corroborating his views, Harvey found it

absolutely necessary to conclude that the blood in the animal body is impelled in a circle, and is in a state of ceaseless motion; that this is the act or function which the heart performs by means of the pulse; and that is the sole and only end of the motion and contraction of the heart.[43]

At first Harvey's book, as he had foreseen, gave rise to much opposition, and Aubrey tells us that he heard Harvey say that "after his book on the 'Circulation of the Blood' came out, he fell mightily in his practice; 'twas believed by the vulgar that he was crack-brained and all the physicians were against his opin-

[40] Cf. *ibid.*, p. 50: "Let us assume . . . . the quantities of blood which the ventricle will contain . . . . two ounces. And one drachm of blood . . . . propelled by the heart at each pulse into the aorta. . . . . In . . . . half an hour, the heart will have made more than one thousand beats. . . . . Multiplying the number of drachms propelled by the number of pulses, we shall have . . . . one thousand drachms of blood . . . . sent from [the heart] into the artery; a larger quantity . . . . than is contained in the whole body. . . . ."

[41] *Ibid.*, p. 48.

[42] *Ibid.*, p. 56 (snake's heart); *ibid.*, p. 69.

[43] *Ibid.*, chap. xiv, p. 71. It should be clearly noticed that Harvey traced the wave of blood from the contracting heart to the arteries and followed back the flow of venous blood to the heart, but did not as yet find the capillary system.

ion."[44] But in the near future Harvey's analysis, supplemented by Aselli's discovery of the lymphatics (1622), and Pecquet's (1651) of the thoracic duct,[45] and by Malpighi's observation of the capillaries (1661),[46] worked a revolution in the science of physiology. Nay more, beyond the realm of physiology it was of the greatest significance in removing the necessity of assuming the three types of spirits in the human system,[47] and went far to eliminate the supernatural as an explanation, substituting in its stead natural law. It showed that accepted notions were fundamentally wrong; that new scientific truths might be attained through the patient, painstaking work of experimentation.

We have to add to the pioneers in the physical and organic sciences in the first decades of the seventeenth century one interesting chemist, the Belgian physician J. B. Van Helmont (1577–1644),[48] a man of the Paracelsus type. He in part fits the age: a careful, exact observer, one who in the spirit of the new physics used measures and weights, and took advantage of the aid of instruments of exact research, and reached his conclusions by accurate quantitative estimates. But in addition, he was a mystic, with such strong leaning toward the supernatural that it is hard to think of him as a contemporary of Harvey, Galileo, and Bacon. Disagreeing with Paracelsus' three and Aristotle's four elements, he recognized only two—air and water. For this he gave the following very interesting proof, which marks the experimenter: He took a pot and earth weighing 200 pounds and a sprout weighing 16 pounds, sprinkled it only with water, whence there developed a tree weighing 169 pounds, pot and earth still weighing 200 pounds. Hence he inferred the tree's growth came

[44] John Aubrey, "Brief Lives," chiefly of contemporaries, set down by John Aubrey, between the years 1669 and 1696; edited from the author's MSS, I, 300.
[45] Foster, op. cit., pp. 47–51.
[46] Ibid., pp. 95 ff.
[47] See above.
[48] Foster, op. cit. (excellent account), pp. 128–44.

only from water, and that the earth was not an element. Helmont was the first to observe that all gases[49] were not identical, and distinguished, for example, carbonic acid from air and vapor. He made extensive studies along lines of fermentation, and these in turn led him along lines of novel physiological speculations. Like Paracelsus, Van Helmont was convinced of the chemical nature of all human processes which he, unlike Paracelsus, resolved into six forms of fermentation; these in turn (here the mystic speaks) he assumed to be directed by a spirit.

This notion of life as a series of chemical processes was the dominating idea of iatrochemists for a century to come, and therefore Van Helmont, in spite of his vagaries, must be reckoned among the great pioneers of the early seventeenth century.

I spoke of two groups of reformers who produced the scientific revolution of 1600–1650, the scientists and philosophic propagandists. It is to the latter group that I now turn: Sir Francis Bacon and René Descartes—two men very dissimilar except in the fact that both were in sympathy with the new movement, and both thought of themselves as experimental scientists; both influenced wide circles whom the direct message of the pure scientist would not have reached; both believed they were originating methods of thought, which, as we have seen, had already been put in practice with remarkable results by active scientific investigators.

Bacon's significance is equally great as iconoclast and as builder; but he will be considered here only in the latter capacity, as the promoter of experimental science. Indeed, he was one of the main forces that brought about the "mutation" from 1600 to 1650. One has but to read the *Advancement of Learning* and the *Novum organum* (Part I) to feel the impact of Bacon's

[49] According to some authorities, the word "gas" was derived by Helmont from *chaos*.

force.⁵⁰ For, as in the case of Luther, it can only be felt at its source. "Other men said the same in whispers, or in learned books written for a circle of select readers. Bacon cried it from the housetops, and invited all to partake. He cried aloud in a language so marvellous, so appealing, so full of pictures, that every sentence carried the full force of conviction. . . . ."⁵¹ Bacon's message is indeed well known. First of all he insisted, by example and precept, upon experiment and observation as the sole means of research, and brought them into a prominence they never had before. The first aphorism of the *Novum organum* strikes the keynote of this message, to be repeated in countless variations: "Man . . . . can do and understand so much, only as he observes in fact or in thought of the course of nature; beyond this he neither knows anything, nor can do anything."⁵² . . . . "Of such observation there will be hardly any . . . . proficience . . . . except there be some allowance for . . . . experiment."⁵³ "For nature like a witness reveals her secrets when put to torture."⁵⁴ Indeed "[deficience in] inventions . . . . of art and sciences . . . . [is] . . . . as if in making an inventory . . . . it should be set down *that there is no ready money*. For . . . . this knowledge . . . . should purchase all the rest."⁵⁵

Let us note here that Bacon ever emphasizes that the ultimate purpose of all scientific knowledge is not the fact per se, but its utilitarian character. "It is not possible to run a course aright

⁵⁰ To quote only one instance of Bacon's effect upon the seventeenth century, Collins, after reading the *Advancement of Learning*, said he found himself "in case to begin studies anew," and that he "lost all time studying before" (J. B. Mullinger, *University of Cambridge*, III, 67).

⁵¹ Thomas Fowler, *Bacon's "Novum organum,"* p. 126: "Ego enim buccinator tantum" (*De augm. scient.*, Book IV).

⁵² Francis Bacon, *Novum organum*, Aph. I.

⁵³ Bacon, *Of the Advancement of Learning*, p. 77.

⁵⁴ Quoted by Fowler, *op. cit.*, p. 127.

⁵⁵ Bacon, *Of the Advancement of Learning*, p. 112.

when the goal itself is not rightly placed. The true and lawful goal of the sciences is none other than this: that human life be endowed with new discoveries and powers....."[56] But if Bacon, in his phrase, "rang the bell to call the wits together" it was for a more specific message, and this indeed he considered his immortal contribution to science. He wanted on every possible topic[57] a collection of all conceivable facts to be made—histories or "calendars, resembling an inventory of the estate of man, containing all inventions which are now extant and whereof man is now possessed."[58] But this collection was not to be in the nature of an enumeration, "for induction which proceeds simply by enumeration is childish,"[59] but according to a fixed and definite method of procedure, viz., the "Baconian instances." These seemed to Bacon as indispensable in the work of the scientist, as the telescope in the study of astronomy: "Neither the naked hand nor the understanding left to itself can effect much. It is by instruments and helps that the work is done, which are as much wanted for the understanding as for the hand....."[60] Such instruments of the understanding were the "instances."[61] The following aphorism clearly shows how different this method is to be from the mere collection of facts:

[56] *Novum organum*, Aph. LXXXI. This, according to Fowler (*op. cit.*, p. 130), was a beneficial influence. "When we recollect the frivolous character of many of the questions which men of most brilliant ability were then in the habit of disputing and the profound misery and discomfort in which the mass of mankind was sunk, we can hardly feel surprise that he [Bacon] advised the application of man's intellectual endowments to the improvement of material conditions."

[57] At the end of *Parasceve* he suggests 130 topics.

[58] *Of the Advancement of Learning*: ". . . . Out of which doth naturally result a note what things are yet held impossible or not invented; which Calendar will be the more . . . . serviceable if to every reputed impossibility, you add what thing is extant which cometh the nearest in degree to that impossibility" (p. 99).

[59] *Novum organum*, Aph. CV.

[60] *Ibid.*, Aph. II.

[61] The text of *ibid.*, e.g., "New Method," Part II.

Those who have handled sciences have been either men of experiment or men of dogmas. The men of experiment are like ants; they only collect and use; the reasoners resemble spiders who make cobwebs out of their own substance. But the bee takes a middle course: it gathers its material from the flowers of the garden . . . . but transforms and digests it by a power of its own. Not unlike this is the true business of philosophy; for it neither relies solely . . . . on the powers of the mind, nor does it take the matter which it gathers from natural history and mechanical experiments and lay it up in the memory whole as it finds it; but lays it up in the understanding altered and digested. Therefore from a closer and purer league between these two faculties, the experimental and rational (such as has never yet been made) much may be hoped.[62]

And such a "league" his "instances" were to supply.[63]

It is clearly brought out in a most interesting dialogue in Spedding's Preface to *Parasceve*[64] that this very thing which Bacon considered the most important part of his work has ultimately not been accepted. But for the purposes of this investigation it is most important to note that even if it did not prove eventually the right method, the seventeenth century accepted it absolutely; the compilation of "histories" was henceforth the ideal of the study of nature.[65] This deeply affected succeeding decades, for this compiling of "histories" was that particular portion of Bacon's message to his century which to us is of greatest importance. Histories as he conceived them could be compiled only by united effort.

Nor was posterity left to surmise in what manner and by what type of organization Bacon advocated such an accumula-

[62] *Novum organum,* Aph. XCV.

[63] Bacon, in his *Sylva Sylvarum* and *History of Winds,* makes collections according to his "instances."

[64] *Op. cit.* (a most enlightening account), pp. 383–403, in John M. Robertson's edition of *Bacon's Works.*

[65] To quote a few instances: Anthony Wood called his work on collected rarities "Britannica Baconica" (Fowler, *op. cit.,* p. 3). The titles of Boyle's works were ever "Histories." Leibniz advocated compiling "calendars" of all facts. (Whether the Baconian instances were used I cannot say.)

RÔLE OF INDIVIDUAL SCIENTISTS 43

tion of facts. It was to be through the instrumentality of learned societies. As Glanvill says: "The great man . . . . formed a society of experimenters of a 'Romantick Model,' but could do no more. His time was not ripe for such performance."[66] This "Romantick Model" is "Salomon's House," a utopian learned society,[67] in the ideal commonwealth Bensalem as described in *New Atlantis*. The "riches" of the house of Salomon consisted in a series of laboratories devoted to all conceivable subjects of experimental research, with facilities of utopian perfection— laboratories beneath the ground, observatories on high towers upon mountain peaks; all apparatus for physiological experiments; botanical and zoölogical experiment stations in the fullest sense of the word; places for dissection, chemico-pharmacological and physical laboratories; special laboratories for the study of heat, of optics, of sound, of engineering problems, all sketched in a completeness which the twentieth century has not reached, but along lines toward which scientific progress has been advancing. All this is put in charge of a hierarchy of scientists, the Merchants of Light, who are to bring news from foreign lands, the Depredators who ransack books for scientific facts, the Mystery Men who collect experiments in the mechanical arts, the Pioneers who try new experiments, the Compilers who tabulate the results, the Dowery men who try to derive practical benefit, the Lamps who direct new experiments, the Inoculators who try these, the Interpreters of nature who "raise . . . . discoveries into greater observations, axioms, aphorisms."

Such is, briefly, Bacon's famous description of the "House of Salomon." It is, as it were, Bacon's last will and testament to his century. It bears to the cause of learned societies the same relation as Marx' "Communist Manifesto" to socialist propaganda. No account is ever given of gatherings of learned men

[66] Jos. Glanvill, *Plus ultra; or, the Progress and Advancement of Knowledge since the Days of Aristotle*.
[67] This description is reprinted in full in the Appendix.

without reference to this "romantick" prototype of societies. To the historical student, to whom the learned societies seem a feature of great importance, there is therefore no personality, except Galileo, so indispensable in a consideration of the progress of science from 1600 to 1650 as Bacon, for he is the veritable apostle of the learned societies.

It is much more difficult—but no less essential—to grasp the significance of Descartes for the cause of experimental science. In several points the French philosopher agrees with Bacon's ideas. To mention first a purely external point, they were both endowed with the extraordinary gift of style, the art of telling their tale in wonderful phrase; while Bacon excels perhaps in brilliancy, there is a note of winning personal appeal, of frankness, of simplicity, in Descartes' writings that establishes closer intimacy between him and his reader. To come to more essential points, Descartes approved wholly and absolutely of Bacon's program of experimentation: "I cannot add anything to Lord Verulanus," he wrote to Mersenne. In the *Discours de la méthode,* Descartes says:

[My work on physics] caused me to see that it is possible to attain knowledge which is very useful in life; and that, instead of that speculative philosophy which is taught in the Schools, we may find a practical philosophy by means of which, knowing the force and action of fire, water, air, the stars, heavens and all other bodies that environ us, as distinctly as we know the different crafts of our artisans, we can in the same way employ them in all those uses to which they are adapted, and thus render ourselves masters and possessors of nature. This is not merely to be desired with a view to the invention of an infinity of arts and crafts which enable us to enjoy without any trouble the fruits of the earth and all good things . . . . but also principally because it brings preservation of health. . . . . As to medicine . . . . all that men know is almost nothing in comparison with what remains to be known . . . . I [decided] to beg all well inclined persons to proceed further by contributing . . . . to experiments and communicating them to the public . . . . in order that the last should commence

RÔLE OF INDIVIDUAL SCIENTISTS 45

where the preceding had left off and thus by joining together the lives and labors of many, we should collectively proceed much further than any one in particular could succeed in doing.[68]

This was surely a Baconian program. Again Descartes says: "I remarked also respecting experiments that they become more necessary the more one is advanced in knowledge."[69] He requests that everybody have cut open for him the heart of a large animal to understand the circulation of blood.[70]

As to learned societies, which are for us such an important part of Bacon's program, Descartes and his followers sympathized with them. Descartes himself was not a member, as no such societies existed in Holland; but he agreed with Queen Christiana's scheme of founding an academy of science, and had drawn up its code and statutes just before his death (February 11, 1650).[71]

Descartes agreed with Bacon in his fundamental disapproval of existing conditions and ideals of learning; indeed, his criticism was even more scathing. Like Bacon, Descartes hated all knowledge which was erudition rather than intelligence. "When one is too curious about things which were practiced in past centuries, one is usually very ignorant about those which are practiced in our time."[72] "There is no more sense in studying Latin and Greek than old Breton or Swiss German," he said; and to Queen Christiana, when he heard she was studying Greek, he pointed to a skeleton and said: "There is my book." He characteristically wrote his *Méthode* in the vernacular.

If I write in French, which is the language of my country, rather than in Latin, ... that is because I hope those who avail themselves only of their natural reason in its purity may be better judges of my opinions than those who believe only in writings of the ancients; and as to those who

[68] Descartes, "The Discourse of Method," Part VI, *Philosophical Works of René Descartes* (ed. E. S. Haldane and G. R. T. Ross), 1911.
[69] *Ibid.*, p. 120.   [70] *Ibid.*, p. 110.
[71] *Encyclopaedia Britannica* (11th ed.), *s.v.* "Descartes."
[72] Descartes, *op. cit.*, p. 84.

unite good sense with study, whom alone I crave for my judges, they will not, I feel sure, be so partial to Latin as to refuse to follow my reasoning because I expound it in a vulgar tongue.[73]

Much more fully than Bacon, he realized the delusion and fraudulent character of "the promises of an alchemist, the predictions of an astrologer (and) the impostures of a magician."[74] He was the foe of conservative reverence for any opinion because it had been adopted, for "there is nothing imaginable so strange or so little credible that it has not been maintained by one philosopher or another."[75] In his travels he saw men entertaining opinions differing from his, and apparently persuaded by just as good reasons—"I concluded that it is much more custom and example that persuade us than any certain knowledge . . . . the voice of the majority does not afford a proof . . . . as truths are more likely to have been discovered by one man than by a nation."[76]

Bacon and Descartes agreed in insisting that nothing was to be admitted as true except what was proved; but at this point they widely disagreed. While to Bacon the sole and only proof was the fact perceptible through sense organs, to Descartes a proof just as real could be derived from human thought and reason. And from this essential point of difference they arrived at widely different conclusions.

It would seem at first thought that Bacon's attitude would be the only one appropriate to the scientists. Yet it must not be forgotten that to Bacon and his followers, in spite of all protestation, there was an island of thought that could not be perceived through sense organs and which had to be accepted by faith; the island where true divinity dwelt. But to Descartes religious speculations ceased to be something apart from other thought. He broke the barrier, coming, it is true, to perfectly orthodox conclusions; but it is the breaking of the barrier which is the

[73] *Ibid.*, pp. 129 f.
[74] *Ibid.*, p. 86.
[75] *Ibid.*, p. 90.
[76] *Ibid.*, p. 91.

RÔLE OF INDIVIDUAL SCIENTISTS 47

important consideration for us at this point. So while Descartes' basis of truth led him ultimately to build up a Weltanschauung little suited to an age of experiment, it led others, perhaps unconsciously, to a rational attitude on religious matters.

Descartes' greatest achievement in pure philosophy was his construction, as it were, of the Euclid of reason.[77] His famous four laws prescribing the mode of all reasoning have been accepted by posterity. The conception of the infinite, he declared, could not have arisen within his finite self, except as it caused itself; hence, he believed, there was a God. Thence Descartes deduced the certainty of human knowledge. God cannot have created man for deception, but all that which man conceives with his pure reason as true is true. What we comprehend clearly and definitely therefore must be true and real.

Let us turn from Descartes the philosopher to Descartes the scientist. He was first and foremost a mathematician, creative and original as perhaps no other man has ever been. This science gave the determining bend to most of his other work. In physics he made fundamental discoveries in the science of optics. Speculations in the field of mechanics occupied his mind to such an extent that in the laws of mechanics he came to see the key to an explanation of the entire universe. This is not the place[78] to give an explanation of Descartes' vortex theory, by which he explained not only all cosmic but all terrestrial phenomena, light, heat, gravitation, lightning, etc. It is well-nigh incredible that a man of mathematical mind, an experimenter, should hand down to posterity—in dangerously plausible terms —a system which so far lacks all mathematical and physical exactness that not one exact magnitude, not one hint at mass, velocity, or space, is to be found. The system was soon accepted among cultured amateurs, and much later by the universities.

[77] Schopenhauer, very sparing of praise, says: "He first induced man to use his own head, for whom hitherto the Bible or Aristotle performed that function."

[78] Rosenberger, *op. cit.* (excellent account), pp. 104–11.

People felt that at one stroke all secrets of the "cosmos" were dispelled. The pleasing and easy demonstrations, mainly by "whirls of water, which pull along everything which comes within their reach," were much more attractive than the difficult laws of Kepler.

It was with the idea of applying mechanical principles to the problems of life that Descartes interested himself in physiology. Of his work in this field Foster says: "Descartes was neither an anatomist nor a physiologist, he studied both as an amateur having a special purpose, to construct out of the current knowledge a physical basis for his philosophical views."[79] His aim was to show that man's structure was similar to that of a machine. For this purpose he wrote the treatise *l'Homme*, which is famous as the first textbook of physiology written in a modern fashion. In a popular way the processes of ingestion of food, digestion, circulation, were explained, with the aim of making it clear that, just as the universe, man and all living things were subject to exact laws of mathematics and physics. In this line of thought he was followed by Steno and Borelli, and became the founder of the iatrophysical school of medicine.

Before leaving Descartes the scientist, we must note one quality which today would be incompatible with a scientific type of mind—his timidity in expressing his scientific conviction. This timidity long prevented him from publishing his *Système du monde,* and when he finally did, his cosmic vortex theory was not given to the world as his theory, but in the following guise:

> Undoubtedly the world was in the beginning created in all its perfection. But as it is best, if we wish to understand the nature of plants or of men, to consider how they may by degrees proceed from seeds rather than how they were created by God in the beginning of the world, so, if we can excogitate some extremely simple and comprehensive principles, out of

[79] Foster, *op. cit.*, p. 58.

RÔLE OF INDIVIDUAL SCIENTISTS   49

which as if they were seeds we can prove that stars and earth and all its visible scene could have originated—although we know full well that they never did originate in this way—we shall expound their nature far better than if we merely described them as they exist.[80]

This timidity—if it can ever be condoned—must be forgiven in the devout Catholic, writing on cosmic matters in the year of Galileo's condemnation. It is but a definite evidence of the baneful effect of the attitude of the Catholic church toward science.

Descartes' cosmic vortex theory and his application of the laws of mechanics to physiology have a significance independent of their truth. The vortex theory was an attempt to deduce all cosmic phenomena from two postulates (extent and motion) with the same cogency as mathematical conclusions are deduced from their hypotheses. It eliminates any need of assuming innate qualities (*qualitates occultae*) in matter.[81] It gave, even if only in a hypothetical form, a mechanistic evolutionary theory of cosmogenesis; and thus in setting, even if not in substance, was as far removed from the account in Genesis as our views today. Similarly in physiology the conception of man as a machine changed the emphasis from soul to body. It put laws of physics and mathematics in place of the mysterious "archai-spirits" of Paracelsus and Van Helmont. And most important, both taken together, as indeed Descartes viewed them, bridge over the chasm separating the organic and inorganic world, and are the earliest attempt at an "einheitliche Weltanschauung."

Descartes stands supreme among those men in the seventeenth century who promoted the mathematical development of the study of science, which came into full bloom in the eighteenth century; but in spite of his own successful work, his bias was against experiment in science. Truths arrived at through

[80] Descartes, *op. cit.*, p. 109.
[81] Such as the circular motion of heavenly, the rectilinear motion of earthly, bodies.

deductive reasoning always remained to him of prime importance. It is most characteristic that in a letter to Mersenne he criticized Galileo for examining experimentally the laws of free fall, without first ascertaining what gravitation is—whether free fall can exist.

Yet even if Descartes furthered but little the cause of experiment directly, he is an indispensable factor in the evolution of science. He brought divinity within the pale of common thought; he widened the compass of natural—in contradistinction to supernatural—laws beyond anything that man had attempted before him. First and foremost he spread scientific inquiry in France and in all Europe through the popular setting of his works. The German phrase, "Er hat Schule gemacht," can hardly be applied with greater propriety to any man. Cartesianism came to be a veritable religion. Even though this included much that experimental science objected to, and finally opposed, it aided the work of the experimenter by methodic doubt, independence of tradition, free thinking, and mechanics as an explanation of natural phenomena. So the Cartesian was an ally in the battle for establishing true knowledge.

After this sketch, however incomplete (for great men such as Gassendi have been omitted), it will become evident why the period of 1600–1650 was called a "mutation," rather than a normal gradual evolution. In these decades, almost simultaneously in the most widely divergent realms of thought, the yoke of tradition was shaken off; there came about a "naissance" (if I may coin the word) of independence from classical inherited thought, a bursting-forth of hidden powers, a change of mental condition from the potential into the kinetic state. As regards science, in 1600 we are in the Middle Ages, in 1650 in modern times.

Turning to the second half of the seventeenth century, we encounter the personality of a man who in every sense is a continuator of Galileo's and Torricelli's work, Guericke, the mayor of

Magdeburg (1602–86).[82] His invention of the air pump furnished an instrument whose wonders were put next to those of the telescope and microscope. It is of great historic interest that he, like many others, was led to experimentation by the contemplation of astronomical matters. He was not a Copernican; to form an independent conception of the universe, he proposed to create experimentally such a medium as the stars move in— "for skill in disputation is of no avail in the realm of science."[83] The experiments[84] he devised for this purpose, epoch making in the annals of the history of science, eventually led him to the construction of the air pump. With this he made his famous experiments, first published to the world by Gaspard Schott (1657).[85] He showed to the astonished Diet of Ratisbonne (1654) that fourteen horses could not separate two metal hemispheres within which a vacuum had been created, but that upon opening a cock to let in air they fell apart. With the pump he drew a column of water to the fourth story of his house, and

[82] Dannemann, *op. cit.*, pp. 166 ff.

[83] Otto Guericke, *Experimenta nova (ut vocantur) Magdeburgica de Vacuo Spatio*, Preface (Ostwald, *Klassiker der exakten Wissenschaften*, Vol. LIX).

[84] Guericke, *op. cit.*, Part III. "I filled a cask with water, made it everywhere air-tight, connected it on the lower side with a metal pump wherewith to draw out the water; I reasoned that as I drew out the water the part of the cask above the water would then be 'empty space.' At the first experiment the cask flew to pieces; I then affixed heavier screws, three men succeeded in pumping out the water, but then a sizzling sound was heard, the air filled the space from which water was drawn. Then I tried putting a smaller cask within the larger, so as to avoid the air rushing into the 'empty space,' and drawing thence the water. Again the sound, now like the twittering of a bird, was heard and lasted for three days, for the wood was porous and let the air through. Therefore I tried a copper sphere instead; first this burst with a loud report. I attributed this to a probable defect in the spherical shape. With the greatest care I had a perfect sphere constructed. Now finally a vacuum was obtained; opening the cock attached to the sphere, the air rushed in with great violence. . . . . In order to show my experiments to the Elector of Brandenburg, I devised the following apparatus, which was easily portable." Then follows a description of the first air pump.

[85] Gaspard Schott, *Mechanica hydraulico-pneumatica*. Through this work Boyle learned of Guericke's experiments.

noted the fluctuations of the column of water and their correspondence to change in atmospheric conditions;[86] he noted that light was propagated in the vacuum, and sound was not;[87] that the life of animals and the flame of a candle could not be sustained in a space without air.[88] All these experiments caused a great sensation and tremendously increased the interest in science.

Guericke's air pump, taken in conjunction with the microscope and telescope and the discovery of the experimental method noted above, give the typical coloring to the mental activity of the decades 1650–90. "It was," says Rosenberger, "in physics the period of the veritable worship of the experiment."[89] Hooke, in 1665, wrote: "We have imperfect senses and imperfect memories, hence imperfect understanding, only the experimental knowledge can rectify these defects."[90] It was even maintained that the experimenter should confine himself to his experiments and avoid the danger of discussing their results.

The great experimenter Boyle thought so little of drawing conclusions from his observations that he left the discovery of the generalization known as "Boyle's Law" to one of his helpers. Everywhere those subjects were chosen where experiment was decisive; the air pump was used and all possible trials made with it; instruments for meteorological observation—barometers, thermometers, hygrometers, anemometers in all shapes, were invented and perfected. Every experimenter dealt with the expansion of bodies through heat, with the problem of boiling and freezing, natural and artificial cold, with color and the study of the spectrum.[91]

We may logically ask: How was this new kind of knowledge spread, and who were the people who first interested themselves in it? In part undoubtedly it was communicated through

[86] Hence he called the instrument *semper vivum*.
[87] Guericke, *op. cit.*, chap. xv.   [88] *Ibid.*, p. 45.
[89] Rosenberger, *op. cit.*, II, 135.
[90] Robert Hooke, *Micrographia: or some Physiological Descriptions of Minute Bodies Made by Magnifying Glasses with Observations and Inquiries Thereupon*, Preface.
[91] Rosenberger, *op. cit.*, II, 135 f.

books; Sir Francis Bacon's works, widely read in England and France, announced loudly that new thoughts were in the air. But the main means of propagation were the discoveries themselves, so intensely novel and startling that as soon as they reached the open-minded man, they would win him forthwith to the cause of experimentation. Who has not been delighted at looking through a telescope or a microscope, or watching a physical or a chemical experiment, even in our own age which accepts the wonders of science as its legitimate heritage? How much more marvelous must these things have seemed to a generation of men to whom they came unexpectedly. The report of a new scientific discovery fills us with admiration and awe, though we are trained to think of science as in a condition of perennial change and continuous growth. How much more amazement must it have caused to minds reared in the idea of the finality and imperturbability of their information. Hooke exclaimed, in commenting on the wonders of the microscope: "In every little particle of matter we behold almost as great a variety of creatures as we were able before to reckon in the Universe."[92] Similarly, Sprat: "We have a greater number of different kinds of things revealed to us than were contained in the visible Universe before."[93] The telescope seemed to have endless possibilities, which even the twentieth century has only in part realized. As Glanvill puts it: "What success and information we may expect from the telescope is romantic and ridiculous to say. . . . . Posterity may find whether the earth moves, and whether the planets are inhabited."[94]

We now come to the question, What class did these new truths and instruments first draw within their spell? It must be emphasized here that experimental science naturally would make its appeal to a vastly larger group of people than that tech-

[92] Hooke, *op. cit.*, Preface.
[93] Thomas Sprat, *The History of the Royal Society of London, for the Improving of Natural Knowledge*, p. 381.
[94] Glanvill, *op. cit.*

nically called the "intellectual class" of the seventeenth century. Scholasticism and humanism from the nature of their teachings had created almost a caste of the learned, and molded the realm of mental activities into an oligarchy or aristocracy; experimental science, on the other hand, stood from its earliest stages for the popularization and hence the democratization of knowledge. While previously the topics and modes of contemplation had been removed from everyday objects and the affairs of men, and confined to regions of speculation barred to most minds, now the subjects and methods of investigation became closely connected with those of homely life. Moreover, the facts of experimental science were of such a nature that they could be comprehended not only by a few highly trained individuals, but by a large number of people of clear mind and comparatively little education. Whereas before, all intellectual activity had been inseparably connected with a mastery of Latin and Greek—an insurmountable barrier to those whose circumstances or inclination prevented them from learning those languages in youth—now the vernacular, at everyone's command, was sufficient linguistic preparation to allow anyone to join in the study of sciences. Before, years of preparation had been necessary to give one the hope, not of adding to, but merely of comprehending the thoughts of those who had gone before them. Now it appeared that the possession of a "faithful hand and an observing eye"[95] put the possibility of sharing in discoveries and valuable work within the reach of vast numbers. Thus experimental science entered the ranks in competition with scholastic learning, and made its strongest appeal not to the erudite university man, wedded to accepted tenets and proud of his place in the oligarchy of the learned, but to the *unzünftigen,* non-professional layman, hitherto excluded from the privileges of mental activity. Indeed, this appeal seems at times to rise to the point of passion. Sprat says (1667): "The love of this science is so strongly roused in

[95] Hooke, *op. cit.*, Preface.

RÔLE OF INDIVIDUAL SCIENTISTS 55

the century in which we live, that it seems there is nothing more in vogue in Europe." This love of science developing among non-university men created the type of the science-loving amateur, so characteristic of the latter half of the seventeenth century. Amateurs in science—"amateurs" in the accepted sense of the word denoting those that "practice their art not as a livelihood, but for the love of it"—were to be found in many places and among many classes of people in the latter part of the seventeenth century, mainly of course in circles which were sufficiently wealthy not to feel the immediate urgency of gaining a livelihood, and had therefore sufficient leisure to follow their inclinations. As these conditions existed, on the one hand, in the larger commercial centers in England and Holland, on the other, in the homes of the nobles and privileged classes, it was in these places that such interest was most conspicuous. An exhaustive study of the amateur scientists of this age—however interesting—is from the nature of the case impossible. I shall merely take up a few individual instances to show how general the interest was and to illustrate the various types of men who became devotees of the new knowledge.

In Italy, Ferdinand and Leopold Medici found a pastime in experimenting, had a laboratory and a collection of instruments, devised experiments, and had their own glass blower.[96] Count Frederigo Cesi,[97] early in the century, was a great lover of science. Count Marsiglio[98] in Bologna was a great experimenter, who gathered men of similar interests about him, and finally bequeathed his home to the university for a laboratory.

In France the Duke of Orleans,[99] brother of Louis XIV, had

[96] Gio Targione Tozzeti, *Atti e memorie inedite dell'Accademia del Cimento*, III, 96 (referred to as *Atti e memorie*).

[97] Domenico Carutti, *Breve storia della Accademia dei Lincei*.

[98] Serafino Mazzetti, *Memorie storiche sopra l'Universita e l'Institute delle Scienze di Bologna* . . . ., pp. 64 ff.

[99] J. Bertrand, *L'Académie des Sciences et les Académicien de 1666 à 1793*, p. 342.

a well-equipped chemical laboratory, loved alchemy, as Saint Simon says, "not to find gold but to amuse himself with curious experiments." He owned a convex lens of great power, in the focus of which metallic gold would melt and become volatilized; he had his own chemists who worked with him. He was besides a great lover of botany,[100] and had the prominent English botanist Morison come to supervise his gardens at Blois.

France is the home of one of the most famous amateurs of all times, Peiresk, the parliamentarian. He was a friend of Galileo,[101] and in frequent correspondence with learned contemporaries.[102] A constant observer of the stars,[103] he bought forty telescopes until he got one good enough to follow the observations of Galileo's *Sidereus nuntius*,[104] and was unhappy because he missed a transit of Mercury.[105] He was equally interested in the shape of snow crystals,[106] fossiled rocks,[107] in fishes and plants.[108] His main business was assisting learned men[109]— so he is depicted by his friend and biographer Gassendi. He was so interested in physiological questions that he had experiments made on a man to test Harvey's discovery.[110] But France, on the whole, produced few experimenting amateurs. Interest there in science often took the form of merely attentive watching of other experimenters' progress, as is seen in the case of Colbert and of Denys de Sallo, the learned founder of the *Journal des sçavans*.[111]

[100] *The Philosophical transactions of the Royal Society of London* . . . . (abridged) I, 341.

[101] Pierre Gassendi, *The Mirrour of true Nobility and Gentility. Being the Life of the Renowned Nicolaus Claudius Fabricius, Lord of Peiresk, Senator of the Parliament at Aix* (Englished by W. Rand; London, 1657), I, 43.

[102] *Ibid.*, p. 36.   [105] *Ibid.*, II, 62.

[103] *Ibid.*, pp. 145–86.   [106] *Ibid.*, I, 210.   [108] *Ibid.*, I, 43.

[104] *Ibid.*, p. 143.   [107] *Ibid.*, II, 46–47.   [109] *Ibid.*, II, 3.

[110] Heinrich Haeser, *Lehrbuch der Geschichte der Medizin*, II, 274.

[111] Louis E. Hatin, *Histoire politique et littéraire de la presse en France*, II, 152.

In England, Charles II is said to have had a "chymist" laboratory and an operator in his palace, and to have frequently visited the laboratories of his friends and discussed scientific questions.[112] Prince Rupert was well known as an amateur. His biographer relates that in 1656 he found new sources of inexhaustible interest in forge and laboratory.[113] After a career as warrior and courtier, he adopted the student's life, and made various inventions, so, for example, he discovered a form of gunpowder ten times the ordinary strength, whereby to blow up rocks in mines and under water, improved locks in firearms, devised a compound of new chemical composition—Prince's metal—and supposedly invented the Prince Rupert glass drop.[114]

There are in England many instances of nobles preferring the study of experimentation to public life. Indeed, it was noted by other nations that the English aristocracy was conspicuously interested in science. As Sprat says: "It [science] has begun to keep best company, refine its fashion and appearance and become the employment of the rich and great, instead of being the subject of men's scorn."[115] "English countryseats removed from the tumult of cities give the best opportunity, and freedom of observation, both of stars and living creatures."[116] First and foremost among science-loving nobles was Lord Robert Boyle, who so completely devoted his time to science that he could hardly be called an amateur, were it not that his chemistry never became a source of livelihood to him, and he never became affiliated with a university. As Boyle is both the highest type of amateur and the foremost figure among English scientists of the time, it will be essential to discuss him somewhat in detail. He is a typical seventeenth-century personality with two distinct

---

[112] Thomas Sprat, *op. cit.*, p. 149.
[113] Eliot Warburton, *Memoirs of Prince Rupert and the Cavaliers*, III, 431 f.
[114] *Ibid.*, II, 433.   [115] Sprat, *op. cit.*, p. 403.
[116] *Ibid.*, p. 405.

natures; on the one hand, he was one of the most skilled experimenters, a pioneer chemist and physicist, who devoted his life and ample means to the acquisition of new knowledge by experiment only, a foe to universal systems such as Descartes';[117] on the other hand, he was a man who was held inconceivably firmly in the clutches of theological speculation and biblical tradition.

It seems fitting in the continuity of history that Boyle should be born in the year of Bacon's death, and that he spent in Florence the winter (1641) in which occurred the death of Galileo, whose work in the line of experiment he was to continue. Boyle, upon coming into his estate at Stallbridge,[118] provided himself with a chemical laboratory.[119] It is noteworthy, however, that much of his time was given to metaphysical and religious speculation, especially to the problem of the reconciliation of science and religion as will be seen from a glance at the titles of some of his essays.[120]

Boyle's conclusions are best expressed, perhaps, in the following sentences from his famous *Some Considerations Touching the Usefulness of Experimental Natural Philosophy:* "Whatever God himself has been pleased to think worthy of making, its fellow creature Man should not think unworthy of knowing"[121] .... "If the omniscient author of nature knew that the

[117] Thomas Birch, *The Works of the Honourable Robert Boyle in five volumes. To which is prefixed the Life of the Author*, I, liv.

[118] *Ibid.*, p. xxx.

[119] *Ibid.*, p. xxxvi. We read of "that great earthen furnace whose conveying hither has taken up so much of my care," also of his limbecks, recipients, and other glasses.

[120] "Excellency of Theology compared with Natural Philosophy (as both are objects of Men's Study) discoursed of in a letter to a friend"; "The Christian Virtuoso showing that by being addicted to experimental Philosophy a Man is rather assisted than indisposed to be a good Christian"; "A Discourse of Things above Reason inquiring whether a philosopher should admit there be any such"; "A Disquisition about final Causes of natural things wherein is inquired whether, and (if at all) with what cautions Naturalists should admit them."

[121] (Edition of 1663), p. 18.

study of his works tends to make men disbelieve his Being or Attributes, he would not have given them so many invitations to study and contemplate Nature. . . . ."[122] Yet he evinced his deep orthodox religiousness by having the Bible translated into Malay for £5,000; by giving £2,000 to missions in America, and leaving £350 in his will to have eight sermons delivered on the truth of Christianity.[123] This was as essentially characteristic of Boyle as his interest in experimentation, and must be emphasized in considering the personality of this scientist.

From his laboratory at Stallbridge, Boyle transferred his residence to Oxford (1654). His biographer Birch relates that in order to prosecute his studies with greater advantage, he chose a private house rather than a college, because he had more room and convenience to make experiments.[124] Provided with all the necessary instruments, and with capable assistants, he experimented incessantly along lines of physics and chemistry which for centuries afterward were conceived as distinct, but have been recently united under the caption "Physical Chemistry." With the aid of his assistant Hooke, he made that famous series of experiments with Guericke's air pump—published as *New Experiments Physico-Mechanical Touching the Spring of Air* (1660),[125] which popularized Guericke's invention and made the instrument more widely known under the name of Boyle's pump.[126]

His chemical experiments are summarized in his book, *The Sceptical Chymist, or Chymico-Physical Doubts and Paradoxes*

[122] *Ibid.*, p. 58.
[123] Poggendorff, *op. cit.*, pp. 467 f.
[124] Birch, *op. cit.*, p. liv. The person with whom he lodged was Cross, the apothecary, a most characteristic choice.
[125] It was in a controversy about these experiments that "Boyle's law" was enunciated.
[126] Rosenberger, *op. cit.*, II, 155. Boyle's physical researches touched the elasticity of water, artificial cold. He was the first to study the theory of colors, thus giving incentive to Hooke's and Newton's researches.

touching experiments whereby Vulgar Spagyrists are wont to endeavor to evince their Salt, Sulphur and Mercury to be true principles of things. This is a conversation in which the "Sceptical Chymist" (being Boyle) refutes three men defining Aristotle's idea of the four elements, and Paracelsus' of three substances composing all matter.[127] Here and in his numberless other chemical works Boyle laid down those chemical principles which made him the father of chemistry, which was now an independent science[128] and no longer the adjunct of alchemy or of medicine. With a comprehensiveness of scientific interest typical of the seventeenth century, Boyle studied anatomy and said of this study: "I have seen in dissection more of variety and contrivance of nature and majesty and wisdom of her author, than all books I ever read in my life could give me convincing notions of."[129]

But the phase of Boyle's work which is the most interesting to us is his advocacy and exposition of right methods of experimentation, shown, especially, in his two essays, *Concerning the unsuccessfulness of Experiments, containing diverse admonitions and observations (chiefly chymical) touching that subject,* in which he stated the correct aim and direction of investigations; and again his constant insistence on the lack of knowledge of essential matters among most men, as in his *Essay of men's Great Ignorance of the Uses of Natural Things, or there is scarce any one thing in Nature whereof the Uses to human Life are yet thoroughly understood.* Indeed it may truly be said that next to Bacon no other individual Englishman did so much as Boyle to advance the cause of experimentation.

Turning now from the professional amateur Boyle to other

[127] Kopp, *op. cit.*, II, 166.
[128] "Postquam chymiae operationes percurrissem, coepi mecum ipse cogitare quanto ad naturalem philosophiam promovendam usui esse possunt . . . ." (*ibid.,* p. 164).
[129] Birch, *op. cit.*, p. xxxiii.

RÔLE OF INDIVIDUAL SCIENTISTS 61

English lovers of science, we meet the personality of the Marquis of Worcester, a wealthy amateur not of pure but of applied science.[130] In his *Century of the names and scantlings of such Inventions as at present I can call to mind to have tried and perfected*, he gave evidences of at least one hundred (as the title says) different experiments he made, among which was the famous one upon which rests his claim of being the inventor of the first steam engine. That he was an enthusiast on the subject of experimental work was shown in his attempt to buy Fauxhall and have it set apart as a resort for artists and mechanics, a depot for models and scientific apparatus, and a place where experiments and trials of profitable inventions could be carried on.[131]

There was also among the English nobles, Lord Bruce,[132] the wealthy Scotch mine-owner, interested in experiments, and Lord Willoughby, who made expeditions to study plants and animals. Then there was Sir Robert Moray, often compared to Peiresc, a military man, one of the privy council of Charles II, interested in science, and owning his own laboratories.[133] There was Sir Mathew Hale (1609–76), lord chief justice of England, "who dedicated no small portion of his time .... to investigations in physics and chemistry, and even to anatomy—and wrote the *Essay touching the Gravitation and Nongravitation of Fluid Bodies*, and *Difficiles Nugae, or Observations touching the Torricellian experiment*.[134] There were Evelyn, the diarist, so interested in science—especially botany—that he proposed to Boyle

[130] Henry Dircks, *The Life, Times and Scientific Labours of the Second Marquis of Worcester.*

[131] Weld, *op. cit.*, I, 53.

[132] *Dictionary of National Biography.*

[133] Aubrey, *op. cit.* "He was a good chymist and assisted his Majestie in his chymicall operations" (II, 82).

[134] *Encyclopaedia Britannica* (11th ed.). It must be noted, however, that he condemned and had executed two women on the charge of witchcraft.

to found with him a sort of scientists' retreat;[135] Kenelm Digby, the well-known cavalier, who kept up a correspondence with learned men,[136] devised cosmetics to enhance his wife's beauty,[137] and wrote a book on the secrets of nature about useful things, especially the sympathetic cure of wounds. There was Samuel Hartlib, "who distinguished himself for great zeal in the improvement of natural knowledge and making it useful for human life,"[138] of whom Evelyn says, "he was a master of innumerable curiosities and very communicative."[139] There was Francis Potter, a man of startling skill in mechanics, according to Aubrey, who "invented and made with his own handes a paire of beame compasses," and originated the "notion of transfusion of blood" (1649).[140]

Also among the rich business men interest in experiment was observable. Notable instances were Sir William Petty,[141] son of a clothier, "who also did [s.c.] dye his owne cloathes," and who later became engaged in that business—famous among writers on economic subjects, and one of the most untiring and ingenious

[135] Weld, *op. cit.*, I, 43–49.

[136] *Commercium epistolicum* (J. Collins, etc.).

[137] Traill, *Social England*, V, 286. (*s.v.*).

[138] Birch, *op. cit.*, p. xxxvii.

[139] *Dictionary of National Biography.*

[140] Aubrey, *op. cit.*, II, 161–70. In the light of recent interest in the question of transfusion, I quote the entire passage (p. 166): "He told me his notion of curing diseases, etc., by transfusion of blood out of one man into another, and that the hint came into his head reflecting on Ovid's story of Medea and Jason, and that this was a matter of ten years ago [hence in 1639]. About a year after he and I went to trye the experiment, but 'twas on a hen and the creature too little and our tooles not good."

[141] *Ibid.*, pp. 139–50. He studied anatomy in Paris; made a model of a double-bottomed vessel with his own hand. Aubrey calls him "a person of an admirable inventive head, and practical parts. He [Sir Petty] hath told me that he hath read but little, that is to say not since 25 aetat; and [thinks] had he read much as some men have, he had not known so much as he does, nor should have made such discoveries and improvements."

RÔLE OF INDIVIDUAL SCIENTISTS 63

of experimenters.[142] Abraham Hill,[143] another business man, was so interested in science that he lived at Gresham College, then the center of experimental knowledge. William Molineux,[144] a rich private citizen residing at Dublin, was the inventor of the hygroscope. Jeremia Horrox (1619-41) and Crabtree, two private citizens, made astronomical observations of great value at Hool, near Liverpool.[145] As Sprat puts it, "The genius of experimenting is so much dispersed that even in this nation, if there were one or two more assemblies settled, there could not be wanting able men enough to carry them on.... all places and corners are busy and warm about this work."[146]

Among the Dutch, the figure of Johann Moritz, count of Nassau Siegen,[147] stands out as a science-loving amateur. He was in supreme command of the Dutch conquest in the New World, and utilized his exceptional position both for his own study and for supplying means of study to others. Swainson praises him in these words:

It is almost inconceivable how this illustrious man, whose life, at this period would appear to have been spent alternately in the camp and the council, could find leisure even to think of science, still less to have prose-

[142] Weld, *op. cit.*, pp. 51-53. We have from his pen an interesting scheme, which shows how he thought that experimental art should be fostered and spread. In his "Advice to Mr. Samuel Hartlib for advancement of some particular part of learning" (1648) he proposed the "establishment of a Gymnasium Mechanicum or a College of Tradesmen; where able mechanics being elected Fellows, might reside rent free." The labours and experiments of these mechanics would be of great value to "active and philosophical heads, out of which to extract that interpretation of nature whereof there is so little and that so bad, yet extant in the world." An institution which "would be as careful to advance the arts as Jesuits are to propagate their religion."

[143] *Dictionary of National Biography, s.v.*

[144] Rosenberger, *op. cit.*, II, 209.

[145] J. H. Mädler, *Geschichte der Himmelskunde*, I, 277.

[146] Sprat, *op. cit.*, p. 71.

[147] *Popular Science Monthly*, LXXXI, 252. William Swainson, "Taxidermy with Biography of Zoölogists," *Cabinet Cyclopedia conducted by Dionysius Lardner*, pp. 259 ff.

cuted it in his closet. Yet the versatility of his mind, and its power of abstraction, was so great that such was actually the fact. He not only patronized and assisted the labors of those whom he had engaged for this purpose, but actually worked himself in describing and drawing the various new animals of Brazil, even in the most arduous periods of his government.

On his expedition to Brazil, he took with him George Marcgrave, the astronomer, geographer, and naturalist; he built in 1639 in Mauritius an astronomical observatory of stone, from which Marcgrave studied the motion of the stars. He had gardens in which large numbers of plants of the country were set out, cages and fishponds for Marcgrave's numerous collections.[148]

The Dutch were throughout the seventeenth and eighteenth centuries famous for their skill in making fine instruments; in a most real sense every grinder of lenses was an amateur scientist. The famous Leeuwenhoek, a linen merchant, self-taught, so little educated that he understood no language but Dutch, gifted however with great manual skill, made a microscope just for his amusement.[149] Gradually he perfected his instrument so that it magnified to 160 diameters, and enabled him to study infusoria. He had so many microscopes that he kept in his investigations one microscope for one or two specimens.[150]

Huygens was an amateur in the same sense as Robert Boyle, giving his whole life to science, but not affiliated with any university.[151] Van Helmont[152] also was a very rich man who had his own laboratory.

In Germany there are famous instances of the interest of amateurs. The rich merchants, the Fuggers, took the scientist

[148] *Ibid.*, quoting John Nienhoff, "Voyages and Travels into Brazil (1640-49)," in John Pinkerton, *A General Collection of the best and most interesting voyages and travels in all parts of the world,* XIV, 710-11.

[149] Haeser, *op. cit.*, II, 296; Carus, *op. cit.*, p. 399.

[150] Jabez Hogg, *The Microscope: its History, Construction and Application,* p. 7.

[151] Dannemann, *op. cit.* (excellent account of Huygens), II, 244 ff.

[152] See above.

RÔLE OF INDIVIDUAL SCIENTISTS 65

L'Ecluse along on their travels.[153] There was Guericke, who, even as mayor of Magdeburg, continued his interest in experimental science, so that when the city was plundered during the Thirty Years' War, he turned to his skill in engineering to gain a livelihood.[154] There was Hevelius (1611–79), the son and heir of a rich brewer in Dantzig, who in 1641 built for himself an observatory, the best equipped of the time, and who ground his own leńses.[155] There was Tschirnhausen, the Saxon duke, who owned three glass factories, and was not only devoted to science, but the originator of famous physical discoveries.[156] There was above all Leibniz,[157] who earned his livelihood as librarian at the court of Hanover, but constantly worked at physical and mechanical problems.

An odd instance of the popular interest taken in science to which the modern word "fad" might apply were anatomical dissections open to the public.[158]

There is another mode of gauging the amateur interest of the nations in experimental sciences other than that of a biographical enumeration of such amateurs. Both in France and in Germany a "popular" work on experiment was widely read, and reached many editions. In 1624 Leurechon published his *Recréation mathématique, composée de plusieurs problèmes plaisants et facétieux en faict d'arithmétique, géométrie, opticque, et d'autres parties de ces belles sciences*. This book was published in seventeen editions in the next year, saw six Dutch, four English translations,[159] and one especially noteworthy, into German

---

[153] Meyer, *Geschichte der Botanik*, IV, 351.
[154] See above.
[155] Wolf, *op. cit.*, p. 321.
[156] Poggendorff, *op. cit.*, pp. 442–45.
[157] See below.
[158] Theodor Puschmann, *A history of medical education from the most remote to the most recent times*, p. 399.
[159] Dr. Friedrich Klee, *Die Geschichte der Physik an der Universität Altdorf bis zum Jahr 1650*, p. 127.

by Schwenter, *Deliciae physico-mathematicae* (1651).[160] It is one of the most instructive books along the lines of this inquiry, and a perusal of its pages gives a clear idea of how much of physics and chemistry may have been within the possession of the interested amateur. We have here the description of experiments made by Schwenter, partly according to Leurechon's direction—all written down avowedly not for study, but for amusement; the experiments are entertaining tricks, not investigations.[161]

The existence of widespread amateur interest in science has been sufficiently illustrated. With this went, in many respects, a readjustment of "values." The calling of an artisan, the profession of the apothecary, rose to a level of respect higher than before. This spirit is summarized in Leibniz' remark:

The charlatans .... and alchymists and .... vagrants are often people of great genius and experience, only they suffer from a "disproportion of genius and judgment" .... surely there is at times in such a man more knowledge won from nature and experience than in a highly respected learned man, who knows how to bring that knowledge he gathered from books, all ready for the market.[162]

[160] The German title is *Mathematische und philosophische Erquickungsstunden darinnen 663 schöne, liebliche und annehmliche Kunststücklein, Aufgaben und Fragen auf der Rechenkunst .... Naturkundigung und anderen Wissenschaften genommen .... allen Kunstliebenden zu Ehren, Nutz, Ergötzung.*

[161] The subject matter of the sections on scientific questions was:
Part V. Optics: telescope. Schwenter enumerates seven colors through a prism, similar to Newton's later theory, and not four, as do all his predecessors
- VI. Catoptrics
- VII. Astronomy and astrology
- VIII. Clocks and magnets. Leurechon's ingenious telegraph from Rome to Paris by means of magnets is described
- IX. Scales and weights
- X. Artificial motions
- XI. Fire and heat. An interesting instrument to keep vessel at steady temperature is described
- XII. Air and wind.
- XIII. Water siphon, waterwheel. A primitive fountain pen is described
- XIV. Chemical arts

[162] Onno Klopp, *Die Werke von Leibniz*, I, 143.

RÔLE OF INDIVIDUAL SCIENTISTS

Enthusiasm for experimentation and the widespread interest it aroused apparently led those devoted to science to enter into more or less formal affiliations. The rich and noble amateur devoted some of his wealth to gathering about him men who would jointly experiment and benefit by this collaboration. The professional scientist would become the center of people who joined him for instruction and whom he needed for assistance. Or at times without any such external stimulus the experimenters would band together.

While unions among people of similar interests and aspirations are natural and have their explanation in man's social instinct, there were cogent reasons why men engaged in scientific pursuit should co-operate. With the development of science along experimental lines, the need for, and expense of, instruments was increasing. Many departments outgrew the stage where bedroom or kitchen laboratories were sufficient; and the interest in science, as we have seen, was not confined to the very wealthy who could afford to equip private laboratories elaborately as did Boyle or the Prince of Orleans. Nor was the laboratory the sole expense. After a new idea was conceived, there was again the great expense of making the desired objects; Kepler never saw a Kepler telescope, only devised it,[163] as he had not the means to procure one; Guericke spent $20,000 upon his instruments.[164] Banding together would seem the only way to obtain means for a laboratory supplied with the necessary instruments, and to obtain the means of making those instruments and experiments which scientific thinkers forecast in their speculations.

How could the average worker afford his own air pump, microscope, not to speak of telescope? How could he ever meet the expense of the improvements constantly suggested, how keep a helper to grind the glasses, a servant to keep the instruments in condition? It cannot be sufficiently emphasized that it was the experimental feature of science which called forth the societies. The mathematician could have worked out his problems in se-

[163] Rosenberger, *op. cit.*, II, 67.   [164] Dannemann, *op. cit.*, II, 167.

clusion; the experimenter needed the laboratory and this in turn could not be supplied under usual circumstances by an individual, but only by a society.

These unions of scientific workers, though there were isolated instances of them before 1650, became the dominating feature of scientific work in the second half of the century, and distinguish it from the previous decades. Not that we do not still find great scientists, whose work shines out quite independently—and in no way depends upon learned societies. Boyle, Huygens, Leibniz, and, above all, Newton, took their place in the history of science along side of Galileo and Harvey. But the fact that their efforts were associated with learned societies during the greater part of their lives and that they co-operated with them, puts their stamp of approval on the forms of scientific work represented by the societies, and allows me to consider them as part of these societies.

It is the main purpose of this dissertation to throw some light upon the manner in which the most prominent of these unions of amateurs crystallizing into learned societies advanced the cause of science. From the nature of the case there must have existed well-nigh innumerable societies, great and small, but we have singled out for closer study: in Italy, the Accademia del Cimento with its precursor the Accademia dei Lincei; in England, the Royal Society; in France, the Académie des Sciences; in Germany, the Academia Naturae Curiosorum, the Collegium Curiosum sive Experimentale and the Berlin Academy. These were the most prominent, but many similar associations, more or less short-lived, existed during the latter half of the century. References are found to numerous societies in Italy[165]—in Venice,[166] in Sienna, the Academia Physico-criti-

[165] For a full list of Italian learned societies, see Büchner, *Academae Sacri Romani Imperii Leopoldino-Carolinae. Naturae Curiosorum historia*, pp. 8 ff.

[166] Vockerodt, *Excercitationes academicae; sive, Commentatio eruditorum Societatibus et varia re litteraria, nec non philologemata socra, auctius et emendatius edita*, p. 20.

RÔLE OF INDIVIDUAL SCIENTISTS 69

ca;[167] in Padua, Academia Constantium;[168] in Naples, Academia Investigantium;[169] in Rome, Academica Physico-Mathematica, which may be identical with the society conducted by Jesuits, mentioned by Poggendorff.[170] Repeated mention is made of an academy in Brescia, Academia Philoexoticorum (1686),[171] of the experimental academy of Bologna,[172] centering about the Duke de Marsiglio. There also existed a society at Aix,[173] one in Denmark at Hafnia, whose publications are of some significance[174] in the annals of medical science. But beyond the fact of their existence, the name of their founder, or of their publications, too little could be ascertained about them to make a study of their activities possible.

In the case of the larger societies, an attempt will be made to sketch their stories with special emphasis upon the way they arose, the type of men who furthered their establishment, helped and joined them; what they did to further the cause of experimental science either in creating laboratory facilities, or in spreading scientific knowledge among scientists or in educating the outside public.

[167] Büchner, *op. cit.*, p. 11.
[168] Vockerodt, *op. cit.*, p. 30.
[169] *Ibid.*, p. 32.
[170] Büchner, *op. cit.*, p. 11; Poggendorff, *op. cit.*, p. 293; Vockerodt, *op. cit.*, p. 32.
[171] This academy published *Acta Nova Academia Philoexoticorum*.
[172] Mazzetti, *op. cit.*, pp. 64 ff.
[173] Vockerodt, *op. cit.*, p. 32.
[174] *Acta Medica et Philosophica Hafniensia*.

PART II
LEARNED SOCIETIES AND JOURNALS

CHAPTER III

ITALIAN SCIENTIFIC SOCIETIES

Italy was the home of the first organized scientific academy, the Accademia del Cimento of Florence (1657–67). It illustrates more perfectly than any other the functions of such societies as centers of the cultivation of experiment. Here nine scientists, supplied with the means of scientific research, gave ten years of united effort to the elaboration of instruments, the acquisition of experimental skill, and the determination of fundamental truths: so completely were their efforts welded together that their work was sent into the world like that of a single individual; so exhaustive were their labors that the book they published became the "Laboratory Manual," so to speak, of the eighteenth century, and their own work and methods the model and inspiration of other learned societies.[1] It will, therefore, be the purpose of this chapter to describe in some detail this Accademia del Cimento, its antecedents, its members, its problems, its achievements, and its failures.

Among the antecedents of the Accademia del Cimento, it is customary to mention the innumerable literary societies which sprang up in Italy, especially the Accademia della Crusca of Florence. But it seems that only two earlier scientific societies in Italy (or for that matter anywhere) could claim any direct influence upon its formation. Giambattista della Porta (1538–1615), the ingenious author of the most popular book on physical magic or magical physics at the end of the sixteenth century, the *Magiae naturalis,* tells in the Preface of an Accademia Curi-

[1] P. van Musschenbroek, *Tentamina Experimentorum Naturalium captorum in Accademia del Cimento,* Preface.

osorum Hominum,[2] which met at his home in Naples and helped him in performing experiments. The condition of membership was that each man had to have made some discovery or communicated a previously unknown fact in natural science. They called themselves "Otiosi," in the style of the literary clubs with odd names then flourishing in Italy. The gathering was referred to as Academia Secretorum Naturae, and judging from the contents of the *Magiae naturalis*, must have tried manifold experiments. For in its twenty books are found, with a singular admixture of alchemy and magic, discussions of magnetism, the *camera obscura*, and of an apparatus which has been acclaimed as a "steam engine" in its most primitive form. The society came to its end very characteristically because Della Porta was accused of meddling with witchcraft, having made a "witches' salve."

More important and better known is the other forerunner of the Accademia del Cimento—the Accademia dei Lincei in Rome (1600–1630). Its device, a lynx with upturned eyes tearing a Cerberus with its claws, was to symbolize the struggle of scientific truth with ignorance. The society was formed in 1601 by Duke Fredrigo Cesi,[3] a man skilful in experimenting, especially interested in the study of bees and plants, fond of collecting natural objects, and in possession of a botanical garden. At his house he met regularly three men interested in the same matters as he—Francesco Stelluti, Anastasio de Filiis, Johannes Eckius—to discuss their studies; these regular meetings and their communications in cipher soon brought on them suspicion of poisoning and incantation, and were therefore broken up.[4] But in 1609 they were reorganized on a larger scale; others joined, such as Della Porta, Peiresc, Galileo (1609), Fabius Colonna, the botanist, until the membership rose to thirty-two.[5]

[2] "Nec domi meae defuit umquam curiosorum hominum academia qui in his vestigandis experiendisque strenuam operam navarent."
[3] Carutti, *Breve storia della Accademia dei Lincei*, pp. 3 ff.
[4] *Ibid.*, p. 12. [5] *Ibid.*, p. 24.

Priests were excluded.[6] The plan was to establish common "scientific, non-monastic monasteries," not only in Rome, but in the four quarters of the globe, for scientific co-operation, a plan interesting as a forerunner of Bacon's "House of Salomon." There was to be a museum, library, printing office, besides optical instruments, machinery, botanical gardens, laboratories— everything that belongs to the study of science. In each house every observation, every discovery, was to be communicated without delay to the head house and all the sister-houses.[7] Their aim was the study of nature;[8] but beyond this, some form of brotherhood was also contemplated. I quote from the *Lynceographum* (1612) in which the "rule of studious life of Lyncean philosophers is laid down:"[9]

The Accademia dei Lincei is a gathering which, according to certain rules and regulations and united friendly councils, directs its labors diligently and seriously to studies as yet little cultivated. Its end is not only to acquire knowledge and wisdom for living rightly and piously, but with voice and writing to reveal them unto men.[10]

While humanistic studies were expressly included in their "Praescriptiones" (1624)[11]—"non neglectis interim amoeniarum musarum et philologiae ornamentis"[12]—as a matter of fact they were not cultivated.

The proceedings of this society were written down as *Gesta Lynceorum*,[13] and have the distinction of being by far the earliest (1609) recorded publication of scientific endeavors by any society. The affiliations of Galileo with the Lincei were very close. He always referred to himself in the dialogues as "Academicus" and added this title to his name in publishing his books. The Academy published two of his works—*Saggiatore* (1612)[14]

[6] *Ibid.*, p. 8.   [7] *Ibid.*, p. 7.   [8] *Ibid.*, p. 26.
[9] "Quo norma studiosae vitae Lynceorum philosophorum exponitur" (*ibid.*, p. 25 n.).
[10] *Ibid.*, p. 7 n.   [12] *Ibid.*, p. 26.
[11] *Ibid.*, p. 63.   [13] *Ibid.*, p. 11.
[14] Strauss, *op. cit.*, Preface, p. xli; Carutti, *op. cit.*, p. 394.

and "On Sun Spots" (1612).[15] After the first condemnation of the Copernican doctrines (1615) a bitter quarrel arose within the band; Lucas Valerius resigned his membership, "because the society, and especially Galileo, upheld forbidden views"; and this must have had grave consequences for Galileo. It was Galileo who made a microscope for the society, and one of its members gave the instrument the name by which it is still called.[16] Stelluti had the distinction of first applying the microscope to the study of zoölogy. The society published his work on bees with the proud inscription: "Stellutus Lynceus Fabrianensis *microscopio* observavit" (Rome 1625).[17] Fabio Colonna published important botanical observations (1624). Then the society issued its most ambitious work, *Thesaurus Mexicanus*, a description of plants and animals of Mexico. With the death of Cesi (1630) the society lost its patrons; and after the condemnation of Galileo (1633) the study of physics and astronomy became too full of dangers to be further pursued by the society. The Academy continued for less than a quarter of a century and finally, in 1657, ceased to exist.[18]

The Accademia del Cimento, i.e., the Academy of Experiment, is closely connected with the Accademia dei Lincei through the personality of Galileo. For, although Galileo was dead at the time of its foundation, he was its spiritual father in a double sense. First, the Cimento was composed for the most part of his own disciples, or of pupils of his two disciples, Viviani and Torricelli, to whom he left the heritage of his mathematical and physical knowledge and experimental skill. Second, it was dedicated in great part to the experimental proof and further elaboration of problems for which Galileo and Torricelli had given theoretical demonstrations.

[15] Strauss, *op. cit.*, p. xxxi; Carutti, *op. cit.*, p. 27.
[16] Carutti, p. 28.
[17] Carus, *op. cit.*, p. 394. [18] *Ibid.*, p. 82.

While Galileo was thus the spiritual father of this Accademia, it was actually called into life by the two Medici brothers, Grand Duke Ferdinand II and Leopold. These, especially Leopold, were of the type of science-loving amateur so characteristic of the time in all ranks of society. They had been pupils of Galileo, Viviani, and Torricelli, and under their inspiration learned to hate scholasticism, to rely upon observation and experimentation. They had a laboratory, and their collection of instruments[19] was shown with pride to all foreigners. We hear, for instance, of Torricelli showing Leopold's instruments to Monconys (1646), and that the collection consisted of Kepler's instruments to measure angles of reflection, various thermometers, eolipiles,[20] instruments for measuring time[21] and the expansive power of powder, and a microscope. The Medici had remarkable collections of animals and wonderful gardens of rare plants. Leopold was so interested in the question of the artificial hatching of eggs that he had two Egyptians come to supervise such an enterprise in his garden.[22] He built incubators, and in contriving instruments to regulate the heat in these, his attention was turned to the study of the thermometer. The dukes instituted regular meteorological observations, where several times daily the temperature and height of the barometer were taken, the direction of winds and the state of the clouds tabulated, these records forming the oldest meteorological tables in existence.[23]

Ferdinand made copious observations on the poison of snakes,[24] on the poison contained in tobacco,[25] on worms gener-

[19] Tozzetti, *Atti e memorie*, I, 153 ff., and II, Part I, 192.
[20] I.e., Heron's ball, rotating through the force of steam.
[21] *Atti e memorie*, II, Part I, 192.
[22] Antinori, *Notizie istoriche relative all'Accademia del Cimento*, p. 34.
[23] Poggendorff, *op. cit.*, p. 383.
[24] *Atti e memorie*, I, 165 ff.   [25] *Ibid.*, p. 169.

ated in vinegar,[26] so it would seem that his interests were biological and chemical. Leopold was a very skilled and original experimenter,[27] and several most ingenious instruments and methods were attributed to him. From 1651 to 1657 we hear of gatherings of scientists at his study (laboratory)[28] alluded to as "Accademia del Principe,"[29] and a list of experiments made there is preserved.[30] The Accademia del Cimento was merely the more formal organization of such meetings and centered entirely about the person of Leopold. The sessions were in a vast room in his residence, next to his library. His instruments—a fine telescope presented to him by Campani, the best then in existence, a microscope given by Divini—and the services of his skilful glass blower, he put at the disposal of the experimenters.[31] The prince defrayed all the expenses. But he was far more than a mere financial patron. He was present at all meetings and hard at work at the experiments,[32] assuming only the humble name and rôle of collaborator. When he traveled, the meetings were interrupted or moved to the place of his temporary sojourn. His wishes and orders were the only constitution or regulation that the Academy had.[33] It was, therefore, natural that when he, in 1667, became cardinal the Academy should have been given up, though there is also a tradition that the Pope exacted its discontinuance as a condition upon which he granted the cardinal's hat to Ferdinand.[34]

Let us turn now to the members of the society:[35] Vincenzo

[26] *Ibid.*, II, Part II, 680.   [30] Antinori, *op. cit.*, p. 93.
[27] *Ibid.*, I, 404.   [31] *Atti e memorie*, I, 380.
[28] *Ibid.*, p. 412.   [32] *Ibid.*, p. 404.
[29] *Ibid.*, p. 376.   [33] *Ibid.*, p. 412.

[34] It is interesting to note that there are many indications that the Cardinal continued to cherish interest in his, once so beloved, natural sciences; he continued in correspondence with the astronomer Hevelius, and Cassini (*Ibid.*, 466; II, Part II, 349) in 1671 surrounded himself with learned men, and had a *History and Meteorology of Etna* written.

[35] Poggendorff, *op. cit.*, pp. 354–72.

ITALIAN SCIENTIFIC SOCIETIES 79

Viviani (1622–1713)[36] was the most typical member; he had been the constant companion of Galileo; so devoted, that he never put his name on a work without calling himself the "last" pupil of Galileo, and never missed an opportunity of rendering homage to him. He was also a pupil of Torricelli, and constructed for him the first barometer. He was a prominent mathematician and won great glory through his reconstruction of Apollonius' works, which tallied with the manuscript found later.[37] He, with Borelli, the second most prominent member, had the responsibility of proposing and showing the experiments.

Borelli (1608–70),[38] by far the most famous member of the Cimento, enjoys a great name in the history of mathematics, astronomy, physics, and especially physiology.[39] In his ten years' affiliation with the Cimento he was most interested in questions of air-pressure,[40] especially in connection with the process of breathing; also in questions of capillarity and experiments with the pendulum. Personally, he was a peevish individual, as is shown by his constant quarrels with Viviani in the Academy, caused by Viviani's above-mentioned work on Appollonius, and in later years by his relationship to Malpighi. Indeed, Antinori sees in these quarrels the reason for Leopold's final dissolution of the Academy. Borelli was the only member who revolted

[36] *Ibid.*, p. 360.
[37] Cantor, *Vorlesungen über Geschichte der Mathematik*, II, 608.
[38] Poggendorff, *op. cit.*, p. 354.
[39] Foster, pp. 61 ff. He was a professor of mathematics, first in Messina, then in Pisa (1665). In his work, *Theoria mediceorum planetarum*, he was one of the closest forerunners of Newton. Of his writings, the best known is the posthumous work *De motu animalium*, in which he reduced in Cartesian fashion all movement to a series of processes subject to exact laws of mechanics. This treatise marks a great advance in physiology, and its conclusions are accepted today. By it Borelli became one of the founders of the Iatrophysical School of physicians, in open opposition to and at that time in great advance of the Iatrochemical School mentioned above.
[40] *Atti e memorie*, II, Part II, 685–93.

against merging his identity with his fellow-workers in the Cimento, and published widely outside its publications.

The other members of the Academy are less well known: Of the two brothers Candido and Paolo del Buono, Candido (1618–76), a pupil of Galileo, was the inventor of a water-clock and several ingenious instruments; Paolo was not much in Florence, but reported from Vienna matters of interest, mainly relating to Leopold's hobby—the hatching of eggs.[41]

The first secretary of the society was Segni;[42] he was soon succeeded by Lorenzo Magalotti (1637–1712),[43] a pupil of Viviani and friend of Robert Boyle. While he was fond of plants and gardening,[44] he seems to have been in no sense a scientist. He had great ability in literary lines, and to him is due the well-finished style in which the Academy finally published its proceedings.

Alesandro Marsili[45] seems of little importance; Antonio Olivia[45] is interesting for a matter of personal history; he was pursued by the Inquisition, and committed suicide in prison (1668) to escape the tortures of the rack. Francesco Redi (1626–94)[45] was a man of standing in the history of zoölogy. He was the physician of the Medici, and founder of the Medical School of Florence,[46] where he insisted on methods of instruction far in advance of his time, even using vivisection. Together with Steno he made microscopic researches; and it is remarkable that in his work he asserted that there was no spontaneous generation, but that all insects come from eggs.[47]

Carlo Renaldini[48] was an engineer, later a professor at the University of Pisa. To him the Academy gave the interesting task of refuting erroneous passages in Aristotle, Pliny, and Gil-

[41] Poggendorff, *op. cit.*, p. 356.
[42] *Atti e memorie*, I, 447.
[43] Poggendorff, *op. cit.*, p. 358.
[44] *Atti e memorie*, III, 112.
[45] Poggendorff, *op. cit.*, p. 359.
[46] *Atti e memorie*, III, 175.
[47] Carus, *op. cit.*, p. 404.
[48] Poggendorff, *op. cit.*, p. 360.

bert.[49] He also worked on the compilation of the *Saggi* of the Academy.[50]

In 1666 these men were joined by the Dane, Steno[51] (1631 and 1686), who had then become the physician of the Medici and who ranks among the first physiologists of the seventeenth century.[52] His personal history is most astonishing. Previously a confirmed Protestant, he later became a Catholic, and ended a life given during its best years to the study of science as bishop of Titiopolis.

To complete here the list of persons connected with the Academy, there were besides these *academici operatori*, corresponding members, among them Michael Angiolo Ricci,[53] who acted as a higher judge of publications, a great friend of Torricelli, interested more in the spread of science than in its promotion, as is evident from his editing the *Giornale dei litterati* in Rome (1668–75).[54] Then there was Honoré Fabri,[55] a French Jesuit, theologian, physicist, astronomer. He seems odd, even as a correspondent of these scientists. He held a position in the Inquisition, and was finally imprisoned for what seems a most conservative statement, namely, that if no proof be found for the

---

[49] *Atti e memorie*, I, 377: "Primo Fascio: Spoglio d'Autori diversi fatti dal Sig. Dott Carlo Rinaldini l'Anno 1656. Questo non era altro que un Quaderno dove alla rinfusa erano notati alcuni passi d'Aristotle, di Plinio, del Gilberto, per verificarli o cofutarli coll' esperienze."

[50] His most important contribution to experimental science, that of suggesting the freezing- and boiling-pcint of water as fixed points for the thermometer scale, falls after the end of the Academy.

[51] Foster, *op. cit.*, p. 106.

[52] Poggendorff, *op. cit.*, p. 369. With Malpighi he was a pioneer in applying a microscope to the study of physiology; with Borelli he was an adherent of iatrophysical ideas. In Florence he turned his thoughts to mineralogy and geology, and in 1669 published a work, *De solida intra solidum*, which, on its rediscovery in 1831, caused the greatest sensation. For he described the earth's crust as "consisting of parallel layers, containing remnants of organic matter."

[53] *Ibid.*, p. 367.

[54] See below, chap. vii.   [55] Poggendorff, *op. cit.*, p. 371.

Copernican system, the church can decide as to its correctness, but if proof be found, the Bible must be differently interpreted. Moreover, the Cimento was in correspondence with Thevenot, the French academician, Boyle, Oldenburg, the secretary of the London Royal Society, and Huygens. Frequently distinguished guests were present at the meetings: Cassini, on his way to Paris in 1665;[56] Professor Sturm of Altdorf,[57] who drew from the Academy the inspiration of forming a similar society in Germany.

And what was the work of these nine *academici operatori* in their daily sessions?

The word *cimento* means "experiment," and this name excellently characterized the society. The Accademia del Cimento did not permit itself to be interested in theoretical physics; its purpose was well expressed in its motto: *"Probando e reprobando."*[58] They desired all efforts to be concentrated upon experimentation, upon the creation of instruments, upon the establishments of standards of measurement and exact methods of research. Only in this way, they conceived, could old errors be refuted, debated matters definitely settled, and new facts accumulated. The meetings of the Academy—always held in private—were most informal. There was only one officer, the secretary; it was his task to prepare the materials and instruments. While to various members particular tasks were allotted;[59] for instance, to make astronomical observations, to write on certain questions and examine certain propositions, the society as a whole worked on the same topic.[60] And just as they experiment-

[56] *Atti e memorie*, I, 249.
[57] *Ibid.*, p. 305; see below, chap. vi.
[58] *Ibid.*, p. 377.   [59] *Ibid.*, p. 412.
[60] Richard Waller, *Essayes of natural experiments made in the Academie del Cimento, under the protection of the most serene Prince Leopold of Tuscany* (written in Italian by its secretary; Englished by R. Waller), p. 417: "It has always been the endeavor in our academy to keep a continued thread of experiments upon what subject so ever they were made."

ed as a unit, their work has come down to posterity, not in signed articles, but as one "anonymous" contribution to the cause of science.

Magalotti in 1667, with the co-operation of the academicians, published an account of their researches under the title: *Saggi di naturali esperienza fatte nell Accademia del Cimento*. A brief analysis[61] of this notable work will best illustrate the ten years' activity of the Academy. It shows the members tirelessly experimenting, constantly trying new methods, not discouraged when their efforts failed; devising elaborate apparatus, heeding details previously neglected, rigorously adhering to their principle —to experiment, never to speculate.

The first pages dealt characteristically with measuring instruments;[62] first of all with the thermometer. There was a description of Galileo's air thermometer, then of an alcohol thermometer[63] used by the Grand Duke of Tuscany in his hatching experiments. Then there followed the description of the four forms of alcohol thermometers used by the Academy, and how they were made by the glass blowers,[64] the critical points still determined by air temperature.[65] Then a thermometer was de-

[61] For a good summary of the contents of the *Saggi*, see Poggendorff, *op. cit.*, pp. 337–403; Rosenberger, *op. cit.*, II, 163–64; E. Gerland und F. Traumüller, *Geschichte der physikalischen Experimentierkunst* (copiously illustrated), pp. 154–77.

[62] Antinori, *op. cit.*, pp. 12–17.

[63] For models of these thermometers, see Deutsches Museum (Munich), Room 21 B C.

[64] The coarser of these thermometers were graduated into fifty divisions; the divisions into ten parts were made carefully with the compass, the subdivisions with the eye. They were marked off with black glass pearls, every tenth with a white pearl. The skilled glass blowers could make several of these thermometers to correspond accurately in their registration of temperature. The next finer instruments were divided into 100°, and one—the most sensitive—was divided into 300°. To attain such exactness the tube was made spiral, and with its three hundred little knobs showed remarkable skill in the maker.

[65] The society, it may be noted, never, in spite of its extensive experimentation with thermometers, conceived the idea of adding a scale, or of deriving the critical points from the freezing- and boiling-point of water.

scribed consisting of balls of various weights floating in the alcohol, and which measured temperature on the principle that as the alcohol expanded and became of lower specific gravity, certain balls would sink to the bottom.[66]

Then hygrometers were taken up.[67] An instrument was described, devised by Ferdinand, measuring the moisture of air by allowing it to be condensed on a vessel filled with ice. The water drops were caught into a graduated cup and the time it took to reach the successive marks served as the measure of the moisture of the air. With this hygrometer they found that southwest winds carry more humidity than north and south winds.[68]

Then the pendulum[69] as an exact time-measuring instrument was studied. "There is need," the *Saggi* said, "of an instrument more exact than the eye or the sound of strokes; for it is difficult to distinguish whether the time is just upon the point marked, or already past when the striking commences." The Academy first devised the bifilary suspension of the pendulum and the method of regulating its vibrations along one plane by wooden guides. The invention of the pendulum clock is here wrongly claimed for Vincenzo, Galileo's son. A clock pendulum beating half-seconds is described.[70]

The first series[71] of the experiments given in the *Saggi* related to "that famous experiment of mercury now spread throughout Europe which first in 1643 offered itself to the thoughts of the ingenious Torricelli."[72] Torricelli's and Roberval's experiments were repeated and objections refuted by experiment. The society very legitimately questioned whether the mercury regis-

[66] For model: Deutsches Museum (Munich), Room 17 K.
[67] Antinori, *op. cit.*, pp. 17-19. For model: Deutsches Museum (Munich), Room 21 N.
[68] Poggendorff, *op. cit.*, p. 388.
[69] Antinori, *op. cit.*, pp. 19-21.
[70] For model: Deutsches Museum (Munich), Room 32 B.
[71] Antinori, *op. cit.*, pp. 23-42.   [72] Waller, *op. cit.*, p. 13.

tered the pressure of all the air or only that beneath the mercury. To answer this query, an elaborate apparatus was devised, careful experiments were made, and the fact definitely established that the pressure of all surrounding air was registered. They repeated in a most original manner, on the tower of Florence, Pascal's experiment[73] to show that atmospheric pressure decreased in greater altitudes.[74] Putting the mercury column to another practical use, they devised an instrument for measuring the specific gravity of liquids by their comparative pressure upon mercury.

In the second series[75] they described instruments for measuring the variation of atmospheric pressure; in the third series,[76] vacuum experiments. For these experiments at first not the air pump but the barometric vacuum was used. For Torricelli's way of testing the effect of the vacuum on insects was to have them crawl through the mercury. This the Academy soon found to be a fallacious method; they therefore adopted Guericke's pump.[77] They tested whether water drops retained their spherical shape in a vacuum, whether heat and cold acted normally, whether Kepler's laws of lenses held good, whether the amber retained its electric, the magnet its magnetic, properties.

[73] See above.

[74] They took a flask with two arms, filled it with mercury, then sealed up one side and carried the flask to the top of the tower; there they saw that the two ends of mercury were no longer on a level, the air in the sealed part pressing heavier than the outside air. They conceived that such an instrument might be of service in measuring the height of the tower, and thought of adding a graduated strip of parchment for that purpose (Gerland-Traumüller, *op. cit.*, p. 161). They were sufficiently exact to realize that the temperature must be measured on the top of the tower and below, and that in case of discrepancy their results would be erroneous; that therefore they must select cloudy days or the time of sunrise for their experiments.

[75] Antinori, *op. cit.*, pp. 43-48.

[76] *Ibid.*, pp. 48-75.

[77] Boyle's air pump, as they wrongly called it, alluding to it as "the pump he described but could not make."

They were much astonished to find that the vacuum did not affect electricity and that capillarity was a phenomenon independent of air-pressure. They tried to devise an apparatus so as not to hear sound in the vacuum, but did not succeed.[78] They repeated "that noble observation of Mr. Boyle of the boiling of warm water"[79] and the action of snow, of smoke, of solutions of pearls and coral in the vacuum. Borelli made experiments which demonstrated the truth of the hitherto unproved assertion, that air was necessary to life:[80] He suddenly removed air from animals by Boyle's pneumatic machine—they fell down dead; if air was instantly renewed with care they could be brought to life again. Such experiments were made on twenty different animals, frogs, birds, fishes, etc.

In the fourth series[81] the much-discussed problem of artificial freezing was taken up; this problem included more than its name suggests, for—say the *Saggi:* "some have thought that cold turns water into the hardest rock crystal and gems of various colors according to different tinctures received from neighboring mineral streams."[82] They attested the expansion caused by freezing by noting that it cracked vessels of most varied substances, and wanted to utilize this force for practical purposes. They added exact tables of their freezing experiments.

The fifth series,[83] "On Natural Cold," contained a famous experiment on radiant heat.[84]

The sixth series[85] dealt with the effects of heat and cold on

[78] They realized that they obtained only a partial vacuum.
[79] Waller, *op. cit.*, p. 57.
[80] Antinori, *op. cit.*, pp. 67–75.
[81] *Ibid.*, pp. 77–105.
[82] Waller, *op. cit.*, p. 69.
[83] Antinori, *op. cit.*, pp. 107–15.
[84] Five hundred pounds of ice were placed so that the cold air struck a concave mirror; a thermometer was placed in the focus of the mirror, and it was found that the temperature was much lower there than in the rest of the room.
[85] Antinori, *op. cit.*, pp. 115–25.

various objects. The scientists had so often to freeze and boil water that they had noticed various peculiarities and decided to examine these very carefully. Thus they put the Aristotelian idea of "antiperistasis" to a test.[86] It seemed to be confirmed by the fact that when an instrument was submerged in ice, the alcohol of the thermometer for a second rose before it fell, and vice versa. The experimenters drew, however, the right conclusion, that this was due to the relative inequality of contraction and expansion of the glass, that antiperistasis was a fallacy, and they even devised an instrument wherewith they demonstrated it. Then they studied the more general problem of expansion of solid bodies through heat; yet never does one meet in their pages any speculation as to what heat itself is—for speculation was prohibited in their meetings.

The seventh series[87] dealt with the question of the compressibility of water. Here they experimented with great pains, and even after many negative results made only the most conservative statement that it was evident that in comparison with air water resisted an "infinitely higher pressure."

The eighth series[88] is most interesting historically, as it refutes experimentally the existence of positive lightness, by showing that smoke and the like do not rise in a vacuum, and therefore their rise under normal atmospheric conditions is only due to their being lighter than air.

The ninth series[89] dealt with magnetism, for instance, with the question whether liquids change the force of magnetism; the tenth series,[90] with amber. Here the *Saggi* relapse into old ways by proving scientific points with quotations from Plato and Plutarch. The eleventh series[91] treated of the changes of colors in

[86] It was assumed by Aristotle and his followers that bodies on being first heated grow cold, and on being cooled grow warm. This was called "antiperistasis."

[87] Antinori, *op. cit.*, pp. 127–30.
[88] *Ibid.*, pp. 131–35.
[89] *Ibid.*, pp. 137–41.
[90] *Ibid.*, pp. 143–47.
[91] *Ibid.*, pp. 149–53.

liquids, e.g., the properties of litmus paper. The twelfth series[92] studied the velocity of sound, repeating the experiments of Gassendi and Mersenne, whereby the velocity was calculated by measuring the difference of time between the lightning and the thunder. The thirteenth series[93] related experiments the Academy devised with which to prove certain statements of Galileo. One of his fundamental theorems—that a projectile thrown horizontally reaches the ground in the same time as one dropped vertically from the same height—was demonstrated by experiment. Galileo's assertion, that the air offers resistance to falling bodies, was ingeniously illustrated by letting similar bodies drop from different heights upon a sheet of tin, showing that the body falling through a greater distance made less of an impact upon the tin. The fourteenth series[94] dealt with a variety of topics, with the question of the weight of air, the effects of cold, whether glass and crystal were penetrable by smell or moisture. The Academy tried to answer experimentally Galileo's question, whether the transmission of light required time. The ingenious experiments gave negative results, because the distance at which they were undertaken was much too small to permit the computation of the velocity of light. Phosphorescence was studied, and lastly, surely under Borelli's inspiration, the digestive processes in animals were investigated.

These are the contents of the famous *Saggi*. The book at first caused no commotion. The first edition was very expensive and could not be bought at bookshops.[95] Italian was not read outside of Italy. Moreover it took some time before it became known abroad. A copy was sent by Magalotti to the Royal Society, March 12, 1668, and Lord Brouncker, seeing how important the experiments were, had the book translated by Richard Waller in 1684.[96] In 1731 P. von Musschenbroek translated it

[92] *Ibid.*, pp. 155–59.
[93] *Ibid.*, pp. 161–64.
[94] *Ibid.*, pp. 165–75.
[95] *Atti e memorie*, I, 459.
[96] Waller, *op. cit.*, Preface.

ITALIAN SCIENTIFIC SOCIETIES 89

into Latin as *Tentamina experimentorum naturalium captorum in Accademia del Cimento,* and this was the edition which formed, as has been said above, the laboratory manual of the following age. In 1755 the *Saggi* were translated into the French *Collection académique.* In 1780 Gio Targione Tozzetti published them again as part of a greater work in three volumes —the *Atti e memorie del Accademia del Cimento.* This contains a most essential supplement to the *Saggi,* for here are given on the basis of diaries of the members, mainly Segni's, a list of many topics investigated by the Academy which did not find their way into the *Saggi.*[97] There are zoölogical researches on fishes, vipers, frogs and worms; very extensive astronomical observations—on Saturn and its rings, the academicians' view of the controversy raging between Huygens and Divini;[98] observations on Jupiter, on the eclipses of the moon and sun and the comets; and (it seems odd to find it) a defense of the Copernican system from the hands of Borelli, in reply to its opponent Ricciolo.[99] The most beautiful reprint of the *Saggi* is that of Antinori in 1841.[100]

As time goes on, the work of the Cimento is more and more appreciated. In books on the history of physics and experimentation, it is generally admitted that it forms an epoch in these sciences. It is the beginning of modern physics. As Poggendorff puts it: "Few bodies have so well fulfilled their aims; the academicians solved the problems they proposed to themselves; we stand today on their shoulders."[101] Rosenberger says of the members of the Cimento:

These first scientific academicians of modern times raised—to the best of their ability—physical questions from the level of doubt to that of certainty. Their measurements were most careful; many measuring instruments owe their origin to them. They showed themselves most cautious thinkers, and did not draw the false conclusions which would seem inevita-

[97] *Ibid.,* II, Part II, 615–798.   [98] *Ibid.,* I, 382.   [99] *Ibid.,* II, Part II, 791.
[100] To this the references have been made.   [101] Poggendorff, *op. cit.,* p. 351.

ble. But their aim was to be exclusively and solely experimental physicists, and therefore they contributed only to the determination of definite facts, but *not* to the development of fertile physical theories. . . . . The value of their results was enhanced by the fact that in the phase of scientific work they adopted, joint effort was most effective; for united forces and united means make for better experimentation, while they would not make for clearer thought or more prolific inventions.[102]

Rosenberger, however, to whom experimentation per se always seems undesirable, laments this limitation of the Academy to experiment, and sees in it an element of exaggeration. He emphasizes that the great Galileo was by no means exclusively an experimenter, and that only his pupils in the first, second, and third generations devoted themselves entirely to it; hence, in his view, the Cimento did not represent the beginning but the end of the most glorious epoch of Italian science. He does not, however, hold the Academy but political and religious influences responsible for what he calls the "one-sidedness" of the Cimento, and suggests that since the discovery of facts was less dangerous than drawing conclusions from investigations, the scientists confined their efforts to experiments only.[103]

It is true that with the extinction of the Accademia del Cimento, Italy's leadership in physics ceased. Her one-time preeminence in matters scientific reduced itself to furnishing Europe with the best telescopes and microscopes, and for a century Italians were the most famous makers of thermometers, and mechanical and optical instruments.[104] But it would be drawing an unwarranted conclusion to say that overemphasis on experimentation caused this. Indeed, this emphasis on experimentation, elaboration of instruments, and experimental methods was a *condicio sine qua non* of scientific progress, and if the Accademia del Cimento stood for this, it certainly merits an important place in the history of experimental science.

[102] Rosenberger, *op. cit.*, II, 164 f.   [103] *Ibid.*, pp. 165 f.

[104] It is interesting that Poggendorff suggests that Italy continued to furnish France with its great scientists.

CHAPTER IV

THE ROYAL SOCIETY

We turn now from the long defunct Accademia del Cimento to the great learned societies of England and France, which continue to exist to this day. While they resembled the Cimento in spirit—since the perfecting of instruments and the cultivation of experimentation was their aim from the beginning, and they continued the investigation of many problems first suggested by the Cimento—they are in many respects very different.

Neither the London Royal Society nor the Académie des Sciences was called into life by a sovereign power, as the Cimento was by the Medici, but arose out of informal spontaneous gatherings of devotees of experimental science, scholars and amateurs. The royal edict did not create them, but only gave a definite and therefore more enduring form to their previous organization. The interests of both these bodies were not so entirely scientific as that of the Cimento; they were well-nigh all-comprehensive, and seemed to subscribe to the motto *"Nihil humanum a me alienum puto."* Side by side with purely scientific problems, there went a consideration of things relating to trade and commerce and manufacture; and it was this phase of their interests which, especially in the first instance, won them the royal patronage. Another essential difference was that in neither the English nor the French society was there that submergence of the individual into corporate effort which was so interesting a feature of the Cimento; personal ambitions were by no means absent in these associations but came into full play, engendering, on the one hand, that healthful emulation which is conducive to progress, but, on the other, giving rise in some cases to intense jealousies which are a blot on the record of their work.

The English and French societies have the special distinction that their form of organization and their methods were destined to become the model not only of the Berlin Academy, which was founded just as the century expired, but of the many learned societies established during the eighteenth and nineteenth centuries; so that in studying their origin we study an important chapter of intellectual history closely connected with important intellectual efforts of our own days.

While the Royal Society and the Académie des Sciences had many elements in common with the Cimento, they nevertheless differed much from one another. The French society was so completely under royal protection that it shows some of the rigidity of a government institution; the English society, supported hardly more than in name by royalty, remained even after its incorporation pre-eminently an informal organization, retaining its amateur features. Here, indeed, all the elements out of which modern science developed can clearly be traced: conscious fidelity to the program sketched by Bacon; close study of Galileo's, Torricelli's, and Harvey's work; a union of the most diverse types of men—business men, divines, nobles, scholars, and physicians; interest in all matters affecting life. Moreover, a recognition of the importance of the inter-communication of ideas gave rise to the establishment of the first periodical scientific publication. Hence a study of the origin and the early years of the English society furnishes a clear idea of the milieu in which natural science was fostered.

Bishop Sprat, in his *History of the London Royal Society*, writes: "If my desires could have prevailed with some excellent friends of mine who engaged me to this work, there should have been no other preface to the *History of the Royal Society* but some of Bacon's writing."[1] He undoubtedly expressed herein the

[1] Thomas Sprat, *History of the Royal Society of London for the Improvement of Natural Knowledge* (1667), p. 35. This is the only contemporaneous account extant.

THE ROYAL SOCIETY

deep truth that Bacon's teachings were influencing the minds and thoughts of all the men who were connected with the founding of this society.

The story of its beginning is as follows: The Royal Society was the result of informal meetings of men more or less learned, but all deeply interested in experimental knowledge. The account of the earliest of such meetings that can directly be connected with the Royal Society is the following from a letter of Wallis:[2]

About the year 1645 while I lived in London (at a time when by our civil wars academical studies were much interrupted in both our Universities) .... I had the opportunity of being acquainted with diverse worthy persons, inquisitive into natural philosophy and other parts of human learning, and particularly of what has been called *New Philosophy* or *Experimental Philosophy*. We did by agreement, diverse of us, meet weekly in London on a certain day, to treat and discourse of such affairs, of which number were Dr. John Wilkins, Dr. Jonathan Goddard, Dr. George Ent, Dr. Glisson, Dr. Merret, Mr. Samuel Foster, then Professor of Astronomy at Gresham College, Mr. Theodore Hank (a German of the Palatinate then resident in London, who I think gave the first occasion, and first suggested those meetings), and many others. These meetings were held some times at Dr. Goddard's lodging in Wood Street .... on occasion of his keeping an operator in his house for grinding glasses for telescopes and microscopes; sometimes at a convenient place in Cheapside and sometimes at Gresham College[3] or some place near adjoining.

Let us consider these men, for a moment, as we shall then the better understand their program of work.

First, as to the scientists: John Wallis (1616–1703)[4] had from his childhood been interested in mathematics, "though mathematics at that time with us was scarce looked on as academical, but rather mechanical—as the business of tradesmen." He had gone to Cambridge and there studied mathematics and

---

[2] Quoted from "Dr. Wallis's Account of Some Passages of His Own Life (1696)," Weld, *A History of the Royal Society*, I, 30 ff.

[3] At Mr. Foster's lodging most likely.

[4] *Dictionary of National Biography, s.v.* John Aubrey, *op. cit.*, II, 281.

medicine, but had acquired a thorough knowledge of contemporaneous continental work in geometry, through independent study. Thus prepared to be a professor of mathematics, he had to leave Cambridge because "that study had died out there, and no career was open to a teacher of that subject,"[5] and eventually he received the Savilian professorship of Oxford (1649–1702). He was the greatest pre-Newtonian English mathematician, and through his important work, *The Arithmetic of Infinitesimals,* paved the way to Newton's discovery of the calculus.

Dr. John Wilkins (1614–72) was the son of a goldsmith, "a very ingeniose man, and had a very mechanicall head. He was much for trying experiments and his head ran much upon the perpetuall motion."[6] Mathematician and astronomer, he had by 1645 written various books on the moon and earth as a planet[7] and was much interested in cryptography. In 1650 he was appointed warden of Wadham College, Oxford.

Samuel Foster[8] was professor of astronomy at Gresham College, and has the distinction of having been expelled temporarily for refusing to kneel at the communion table. Most of his work was on astronomical instruments.

As to the physicians, they were all associated with the Royal College of Physicians. Dr. Goddard (1617–74), Cromwell's physician, was professor of medicine at Gresham College; there he "lived and had his laboratory for Chymistrie," also for pharmacological purposes. It is known that he sold drops to Charles II for two thousand pounds. In his laboratory he made a number of experiments for the Royal Society. "They made him their drudge, for when any curious experiment was to be done they would lay the task on him."[9] He was the chemist of this com-

[5] James Bass Mullinger, *The University of Cambridge,* III, 462.
[6] Aubrey, *op. cit.,* II, 299.
[7] Cantor, *op. cit.,* III, 339.
[8] *Dictionary of National Biography, s.v.*   [9] Aubrey, *op. cit.,* I, 268.

pany. Dr. Ent (d. 1679)[10] and Dr. Glisson[10a] were friends of Harvey, the latter the first to teach the circulation of the blood. They represented thus the physiologists and anatomists of the society. Dr. Merret,[10a] another friend of Harvey, made large collections of British plants and represented the botanists. Hank[10a] seems to have had no specialized interests.

Returning from these biographical considerations to Wallis' letter,[11] we read:

Our business was (precluding matters of theology and state affairs) to discourse and consider Philosophical Enquiries and such as related thereunto; as Physick, Anatomy, Geometry, Astronomy, Navigation, Staticks, Magneticks, Chymicks, Mechanicks, and natural Experiments; with the state of these studies, as then cultivated at home and abroad. We then discoursed of the circulation of the blood, the valves in the veins, the *venae lacteae,* the lymphatick vessels, the Copernican hypothesis, the nature of comets and new stars, the satellites of Jupiter, the oval shape (as it then appeared) of Saturn, the spots in the sun, and its turning on its own axis, the inequalities and selenography of the Moon, the several phases of Venus and Mercury, the improvement of telescopes, and grinding of glasses for that purpose, the weight of air, the possibility, or impossibility of vacuities, and nature's abhorrence thereof, the Torricellian experiment in quick-silver, the descent of heavy bodies, and the degrees of acceleration therein; and divers other things of like nature. Some of which were then but new discoveries, and others not so generally known and embraced, as now they are, with other things appertaining to what hath been called the *New Philosophy,* which from the times of Galileo at Florence, and Sir Francis Bacon in England, hath been much cultivated in Italy, France, Germany, and other parts abroad, as well as with us in England.

The letter goes on to speak of the continuation of these London meetings in Oxford:[12]

---

[10] Haeser, *op. cit.,* II, 252. Dr. Ent published Harvey's book, *De generatione animalium.*

[10a] *Dictionary of National Biography, s.v.*   [11] Weld, *op. cit.,* I, 31 f.

[12] It is a debated question whether meetings of the same type previously existed independently in Oxford about the person of Sir William Petty, and were only strengthened by the London contingent. The correspondence of Boyle and Hartlib, 1646-47 (see Birch, *Life of Robert Boyle,* I, xxix), contains frequent al-

About the year 1648-49 some of our company being removed to Oxford, our company divided. Those in London continued to meet there as before (and we with them, when we had occasion to be there); and those of us at Oxford with Dr. Ward (since Bishop of Salisbury), Dr. Ralph Bathurst (now President of Trinity College in Oxford), Dr. Petty, Dr. Willis (then an eminent physician in Oxford), and divers others, continued such meetings in Oxford, and brought those studies into fashion there; meeting first at Dr. Petty's lodgings (in an apothecarie's house), because of the convenience of inspecting drugs and the like, as there was occasion; and after his remove to Ireland, at the lodgings of Dr. Wilkins, then Warden of Wadham College, and after his removal to Cambridge, at the lodgings at the Honourable Mr. Robert Boyle, then resident for divers years in Oxford.[13]

We have the testimony of Aubrey,[14] Evelyn,[15] and others that through Wilkins' influence a veritable deluge of scientific interest invaded Wadham College. There resided Seth Ward, Christopher Wren, and Rooke, all enthusiasts of the new study. Seth Ward (1617–88) was a fellow-student of Wallis in mathematics. He was expelled from his Alma Mater for refusing to subscribe to the League and Covenant;[16] and was prevailed

---

lusions to an "Invisible College" and points beyond doubt to the existence of such a gathering. Moreover, Sprat, *op. cit.*, p. 53 (a great Oxford admirer), writing his history in 1667, mentions only such Oxford meetings as antecedents of the society: "The University had at that time many members of its own, who had begun a free way of reason and was frequented by some gentlemen whom the misfortunes of the kingdom drew thither." Yet, according to Wallis' letter which we are quoting, the Oxford meetings are but the London meetings transferred.

[13] Weld, *op. cit.*, I, 33.

[14] Aubrey, *op. cit.*, II, 301, says of Wilkins: "He was the principall reviver of experimentall philosophy (*secundum mentem domini Baconi*) at Oxford, where he had weekely an experimentall philosophicall clubbe . . . . which was the incunabula of the Royall society."

[15] J. Wells, *Oxford University College Histories, Wadham*, p. 73. Evelyn says of Wilkins: "He has in his lodgings and gallery a variety of shadows, dials, perspective and many other artificial mathematical and magical curiosities, a waywiser, a thermometer, a monstrous magnet, most of them of his own or of Chr. Wren."

[16] Aubrey, *op. cit.*, II, 284.

upon to take the Savilian professorship of astronomy.[17] He was so enthusiastic that he proferred to all comers gratuitous instruction. Christopher Wren (1631–1723), mainly remembered by posterity for his architectural abilities, was one of those rare geniuses whose mind seemed to compass apparently the most diverse subjects of human inquiry. At Oxford he ranked high in his knowledge of anatomical science;[18] he was the successor of Seth Ward as Savilian professor of astronomy, and constantly engaged in experiments at his chambers in Wadham with Wilkins. Prominent in this group of enthusiasts was Laurence Rooke (1623–62), so devoted a pupil of Seth Ward that he made him the sole beneficiary of his will.[19] He acted as Boyle's assistant in his chemical operations until in 1652 he was called to Gresham College, first as professor of astronomy and later as professor of geometry. He was called the "greatest man in England for solid learning, not excepting botany, music and divinity,"[20] and Aubrey relates, "he took his sickness of which he dyed by setting up often for astronomical observations."[21] Dr. Bathurst, who was mentioned in Wallis' letter, lived at Worcester College, which seems at this time to have been the center of some scientific interest.[22] Dr. Thomas Willis (1621–75) was a

[17] Mullinger, op. cit., III, 315, relates that he was persuaded by the following argument: "If you refuse, they'll give it to some cobbler, who never heard the name of Euclid and mathematics, who will snap at it eagerly for salaries' sake."

[18] Weld, op. cit., I, 273. His ability as a demonstrator, and his attainments in anatomy generally, were acknowledged with praise by Dr. Willis in his *Treatise on the Brain*, for which Wren made all the drawings.

[19] *Dictionary of National Biography*, s.v.   [20] *Ibid.*

[21] Aubrey, op. cit., II, 204.

[22] C. H. Daniel, W. R. Barker, *Oxford University College Histories, Worcester*, p. 125. There lived Thomas Allen: "The vulgar did verily believe him to be a conjurer. He had a great many mathematical instruments and glasses in his chamber which did also confirm the ignorant in their opinion; and his servitor . . . . would tell them that sometimes he would meet spirits coming up his stairs. . . . . There went from hand to hand a volume of letters concerning chemical and magical secrets."

student of medicine and the natural sciences, especially chemistry.[23] In 1660 he held the Sedleian chair of natural philosophy. He was one of the foremost English physicians and the author of an epoch-making work on the brain.[24]

The union of these scientists bore the name Philosophical Society of Oxford, and its regulations are preserved.[25] Their weekly meetings are thus characterized by Sprat:

> Their proceedings were rather by action than discourse, chiefly attending some particular trials in Chymistry or Mechanicks; they had no rules nor method fixed. Their intention was more to communicate to each other their discoveries which they could make in so narrow a compass, than united, constant and regular disquisitions.[26]

Birch relates that these men, "being satisfied that there was no certain way of arriving at competent knowledge unless they made a variety of experiments upon natural bodies in order to discover what phenomena they would produce, pursued that method by themselves with great industry and communicated it to each other."[27] Some interesting hints as to their work are printed among the notes of Wood's *Diary* (1659) under the caption "The Royall Societe at Oxon, and of Chemistry":

> They did in Clerk's house, an apothecary, exercise themselves in some chemical extracts, ... in so much that severall Scholars had private Elaboratories and did perform those things, which the memory of man

[23] For an excellent account of Willis, see Haeser, *op. cit.*, II, 382. He understood the great significance of chemistry in the study of physiology. In opposition to current views he spoke of an animal "psyche" (Aubrey, *op. cit.*, II, 303 f.).

[24] In this he described what is still known in the anatomy of the brain as "the circle of Willis." Something has been said above in chap. ii about Dr. Petty, the scientific business man, and Robert Boyle, the foremost scientific amateur of England during the century.

[25] Quoted by Weld, *op. cit.*, I, 33 f., from MS in Ashmolean Museum. I cite one interesting regulation: "If any man does not duly on the day appointed perform such exercise or bring in such experiment as shall be appointed for that day, or in case of necessity provide that the course be supplied by another, he shall forfeit two shillings six pence and shall perform his task notwithstanding. . . . ."

[26] Sprat, *op. cit.*, p. 56.

[27] Birch, *op. cit.*, I, xxxiii.

could not reach. But the one man that did publickly teach it to the scholars was Peter Stahl[28] .... brought to Oxon by Mr. R. Boyle and by him settled in the same house wherein he lived, ... where, continuing a year or two and taking to him disciples, [he] in time translated himself to a tenement near it, and then to an ancient hall called Rain Inn, in the old refectory of which he erected an elaboratorie and taught several classes. Among such that he taught were Dr. John Wallis, Mr. Christopher Wren, Mr. Nathaniel Crew, Dr. Ralph Bathurst, Dr. Richard Lower.[29]

This laboratory of Stahl's seems the one sometimes referred to as fitted up for the Oxford society.[30]

By 1665 so many men of the Oxford group moved away that Stahl, for want of disciples, left for London to assume the position of operator of the Royal Society.[31] Indeed ever since 1658 many of the Oxford body had returned to London,[32] meeting

[28] Not to be confused with the great Stahl.

[29] Andrew Clark, *The life and times of Anthony Wood, antiquary of Oxford (1632–1695), described by himself*, I, 290. And another account (1663) by the same pen (*ibid.*, p. 473): "There began a course of chemistry of Stahl. The club consisted of ten, among whom was John Locke."

[30] Wordsworth, *Scholae academicae—Some account of studies at English universities in the eighteenth century*, p. 176. Weld, *op. cit.*, I, 43 ff.: How much the need of a common meeting ground was felt is evidenced by a letter written September 1, 1659, by Evelyn to the rich bachelor Boyle, for erecting a "Philosophic-Mathematical College." He proposed to purchase property within twenty-five miles of London, have apartments or cells for members of the society "somewhat after the manner of the Carthusians." "There should be an elaboratory with a repository for rarities and things of nature, an aviary, dove house, physick garden, of four hundred pounds running expenses, one hundred forty-five pounds to be employed for books, instruments, drugs and trials. Every person of the society shall render public account of his studies weekly if thought fit, and especially shall be recommended the promotion of experimental knowledge as the principal end of the institution." It is thus a scientific kind of monastery to which Evelyn wants to retire with "such a person as Mr. Boyle, who is alone a society of all that were desirable to a consummate felicity."

[31] Clark, *op. cit.*, I, 290.

[32] Weld, *op. cit.*, I, 35. This Oxford society under the name of the Oxford Philosophical Society continued to exist until 1690, and was in correspondence and co-operation with the London Royal Society. It tried to enlist the learned bodies in the cause of science. For instance, in a letter to the heads of universities, they wrote (quoted by Weld): "We would by no means be thought to

now at lectures in Gresham College. This institution is so intimately connected with the early history of the English society that a few words about it must be said here. Sir Gresham in his will (1575) had left valuable property to the citizens of London to provide them with a college in his former mansion. There seven professors were to live in commodious apartments, and were to deliver a daily lecture to citizens of London on divinity, astronomy, music, geometry, law, physics, and rhetoric—a remarkable instance, indeed, of interest in the spread of science, and in an attempt to reach the unlettered people.[33] Two of the men of the Oxford society, Rooke and Wren, had been appointed to Gresham professorships. It was at their lectures that many of the previous Oxford group, now residents of London, met. Sprat reports: "Here joined with them several eminent persons, Lord Brouncker, Lord Brereton, Sir Paul Neil, Mr. John Evelyn, Mr. Henshaw, Dr. Slingesby, Dr. Timothy Clark, Dr. Ent, Mr. Ball, Mr. Hill, Dr. Crone and divers other gentlemen whose inclination lay in the same way."[34]

We are now close to the body that was to be incorporated as the Royal Society, and therefore naturally ask what manner of men these were. Lord Brouncker (1620–84)[35] had translated Descartes' *Musical Compendium;* his main interest lay in mathematics, to which he contributed the theory of continued fractions and the quadrature of the hyperbola;[36] he, with Evelyn, often discussed scientific questions with Charles II.

---

slight or undervalue the philosophy of Aristotle . . . . but we do not think [nor did he think] that he had so exhausted the stock of knowledge that there would be nothing left for inquiry of after times, as neither we of this age hope to find out so much but that there will be much left for those that come after us." The letter proceeds earnestly to request the assistance of members of colleges, etc., towards the great work of advancing scientific knowledge. The fact that the society ceased in 1690 is the best comment on the little success their efforts met with. (See below, chap. viii.)

[33] Weld, *op. cit.*, I, 80 ff.
[34] Sprat, *op. cit.*, p. 57.
[35] Aubrey, *op. cit.*, I, 129.
[36] Cantor, *op. cit.*, III, 55 ff.

THE ROYAL SOCIETY

Evelyn was the well-known diarist; Henshaw,[37] a lawyer and the King's undersecretary, was interested in chemical matters; Mr. Ball[37] was an astronomer; Dr. Clarke[37] a physician, who conducted some dissection before Charles II and encouraged the royal interest in this direction. Mr. Hill[38] was a business man who eventually became commissioner of trade.[39]

These meetings at Gresham College were rudely interrupted when the building was converted into barracks for the soldiers, but with the "wonderful pacifick year" (1660) of the Restoration, they were revived and attended by an increased number, and on November 28, 1660, those in attendance constituted themselves into a definite association. With an account of this event the first journal-book of the Royal Society opens:

Memorandum November 28, 1660. These persons following, according to the usual custom of most of them, met together at Gresham College, to hear Mr. Wren's lecture.[40] And after the lecture was ended they did according to usual manner withdraw for mutual converse. Where amongst other matters that were discoursed of, something was offered about a design of founding a college for the promoting of *Physico-mathematical Experimental Learning;* and because they had these frequent occasions of meeting with one another, it was proposed that some course might be thought of, to improve this meeting to a more regular way of debating things, and according to the manner in other countrys where there were voluntary associations of men in academies for the advancement of various parts of learning, so they might do something answerable here for the promotion of experimental philosophy.[41]

They agreed to meet Wednesday at Mr. Rooke's chamber at Gresham College, and in vacation at Mr. Ball's chamber. Dr. Wilkins was appointed chairman, and a list of forty-one persons,

[37] *Dictionary of National Biography, s.v.*
[38] See above, chap. ii.
[39] About Lord Brereton (1631–80) and Paul Neil I find nothing.
[40] Viz., Lord Brouncker, Mr. Boyle, Mr. Bruce, Sir Robert Moray, Sir Paul Neil, Dr. Wilkins, Dr. Goddard, Dr. Petty, Mr. Ball, Mr. Rooke, Mr. Wren, Mr. Hill.
[41] Weld, *op. cit.*, I, 65.

"judged willing and fit to join with them in their design," was drawn up.

On the Wednesday following the opening entry of the journal-book, Sir Robert Moray, who was called the "soul of the meetings" and represented the connecting link between the King and the scientists, brought word that the King approved of their design, and from thenceforth the journal reports weekly meetings. December 5, 1660, a formal agreement was signed by the original members and seventy-three others, "to consult and debate concerning the promoting of experimental learning." Informal rules were drawn up.[42] Gresham College after some debate was designated as a permanent place of meetings. The young society had not met many months before a poem was written, probably by William Glanvill, "In praise of the choice company of Philosophers and Witts who meet Wednesdays weekly at Gresham College":

>   At Gresham College a learned nott
>   Unparalleled designs have layed
>   To make themselves a corporation
>   And know all things by demonstration.
>
>   These are not men of common mould,
>   They covet fame but condemn gold.
>   The College Gresham shall hereafter
>   Be the whole world's University,
>
>   Oxford and Cambridge are our laughter;
>   Their learning is but pedantry.
>   These new Collegiates do assure us
>   Aristotle's an ass to Epicurus.
>   —W. G.[43]

[42] The number of members was restricted to fifty-five, but any baron, any Fellow of the Royal College of Physicians, and public professors of mathematics, physics, and natural philosophy could join. There was as yet no curator—merely a president, treasurer, register, amanuensis (with a salary of forty pounds), and operator (with a salary of four pounds); members were to sit with the registrar taking notes.

[43] Weld, *op. cit.*, I, 79 n.

The formative period of the society dates from November 28, 1660, to July 15, 1662, when the royal charter was issued, for the lines were then laid out along which its future work proceeded. The question of greatest interest, How was experimentation really carried on? is hard to answer. The Gresham professors formed, as it were, a "committee on experiment," and "every man of the company was desired to bring in such experiment as he shall think most fit for advancement of the general design of the company." We hear[44] that Mr. Boyle showed his experiment of the air; Dr. Clark, the injection of liquid into the veins of animals. So we know at least of two types of experiments of the greatest importance. As to obtaining a laboratory, a committee was formed to consider all sorts of tools, instruments and "glasses for perspectives" for the society.[45] As to the systematic preparation for the meetings it is recorded that when Mr. Boyle presented his book concerning glass tubes, it was ordered that each member have one in order to discourse on it at the next meeting.

Apart from laboratory work, another line of investigation was at once started, namely, to procure authentic information on the natural history and physical condition of foreign countries, respecting which great ignorance prevailed.[46] Interest in commerce and trade was evinced. *A Philosophy, i.e., History of*

[44] *Ibid.*, pp. 95 ff. *passim.*

[45] "An Essay furnace was ordered to be built and accurate beam [scale] provided for the use of the society." Mr. Boyle presented the company with an air pump.

[46] Boyle and Brouncker drew up twenty-two "questions" to send to Teneriff, e.g., "Try the mercury experiment at the top and several other ascents of the mountain. Take by instrument (with what exactness may be) the true altitude of every place where experiment is made. Observe the temperature of the air by a weather glass, and as to moisture and dryness with a hydroscope. . . . . Try by an hour glass whether a pendulum clock goeth faster or slower on top of the hill than below." The East India Company agent was given a series of questions with the same intent. See Weld, *op. cit.*, I, 117; also Sprat, *op. cit.*, pp. 200 ff.; for answers to inquiries about Teneriff.

Shipping and Clothing and Dyeing, was allotted to Sir W. Petty, subjects closely related to his past business interests.[47] The rule that members should present their books to the society was established. Formal correspondence with foreign learned bodies was opened, for instance, with Duke Leopold of the Accademia del Cimento, and with the group at Montmort's house soon to be established as the Académie des Sciences.[48]

The various papers submitted during this year show a somewhat astonishing absence of criticism. We must not forget that, after all, the majority of the members were of average mentality. President Sir Robert Moray handed in a paper on *Barnacles* reporting that when he was in the western island of Scotland he saw a multitude of little shells adhering to trees, having within them little perfectly shaped birds; yet he never saw any of the birds alive. There were papers on the production of young vipers from the powdered liver and lungs of vipers, and on magnetic and sympathetic cures.[49] These give the best evidence of how much a scientific society was needed.

The King, who had always—presumably through Sir Robert Moray—shown great interest in the work of the savants, and sent them "rarities," on July 15, 1662, established them as a Royal Society by a charter—called by some the only wise act of Charles II. The preamble of the charter reads as follows:

And whereas we are informed that a competent number of persons of eminent learning, ingenuity and honour, concording in their inclinations and studies towards this employment, have for some time accustomed themselves to meet weekly and orderly to confer about the hidden causes of things, with a design to establish certain and correct uncertain theories in philosophy, and by their labour in the disquisition of nature to prove themselves real benefactors to mankind; and that they have already made a considerable progress by divers useful and remarkable discoveries, inventions and experiments in the improvement of Mathematics, Mechanics, As-

[47] See above, chap. ii.
[48] See below, chap. v.
[49] Weld, *op. cit.*, I, 107 ff.

tronomy, Navigation, Physics and Chemistry, we have determined to grant our Royal favor, patronage and all due encouragement to this illustrious assembly, and so beneficial and laudable an enterprise.[50]

The charter opened with the following interesting statement:

> We have long and fully resolved with ourselves to extend not only the boundaries of Empire but also the very arts and sciences. Therefore we look with favor upon all forms of learning, but with particular grace we encourage philosophical studies, especially those which by actual experiments attempt either to shape out a new philosophy or to perfect the old. In order therefore that such studies which have not hitherto been sufficiently brilliant in any part of the world may shine conspicuously amongst our people and that at length the whole world of letters may always recognize us not only as the Defender of the Faith, but also as the universal lover and patron of every kind of truth . . . . know ye that we . . . . have ordained . . . . there shall be a society consisting of a President, Council and Fellows, which shall be called and named The Royal Society. . . . . The Council shall consist of twenty-one persons, of whom we will the President to be always one. . . . . And that all and singular other persons who within one month . . . . shall be received and admitted by the President and Council . . . . shall be called and named Fellows of the Royal Society, whom, the more eminently they are distinguished for the study of every kind of learning and good letters, the more ardently they desire to promote the honour, studies and advantage of this society . . . . the more we wish them to be especially deemed fitting and worthy of being admitted into the number of Fellows of the same society.[51]

The first president and council were appointed by the King. The president was Lord Brouncker; the council: Robert Moray, Robert Boyle, William Brereton, Kenelm Digby, Paul Neil, Henry Slingesby, William Petty, Professor John Wallis, Dr. Timothy Clark, Professor John Wilkins, Dr. George Ent, William Aerskin, Dr. Jonathan Goddard, Dr. Christopher Wren, Saville, William Ball, Mathew Wren, Evelyn, Thomas Henshaw, Dudley Palmer, and Henry Oldenburg. The society was to have a treasurer—the first appointee being William Ball—and two

[50] *Ibid.*, p. 121. [51] *Record of the Royal Society of London* (1901), pp. 31 ff.

secretaries—these posts being assigned to Wilkins and Oldenburg.[52] They were to meet in the college or any public place or hall in London, or within ten miles, and make laws, statutes, and ordinances relating to the affairs of the society.[53] They could appoint printers and engravers, and obtained the right to demand, receive, and anatomize bodies of executed criminals, the same as the College of Physicians and Corporation of Surgeons in London.[54] And further the charter continued: "We have given .... that they .... enjoy mutual intelligence and knowledge with .... strangers and foreigners .... without any molestation, interruption or disturbance .... provided [this be] in matters or things philosophical, mathematical and mechanical."[54] That the possibility of the function of teaching was considered is evident from the fact that they obtained a license to build a college in London—or within ten miles.[55]

Thus, in a few years the "Invisible College" of Boyle was incorporated by royal charter, and from a few philosophers and lovers of science, meeting here and there as times permitted, there grew a society that soon acquired a stability which two and a half centuries have not weakened.[56]

The charter made no financial provision whatsoever.[57] This was the great difference between the English and the French societies, and indeed, the Accademia del Cimento. While these were in good financial condition, the Royal Society was for years in a

[52] Oldenburg was destined to accomplish invaluable work for the society. A native of Bremen, professor at the University of Dorpat, he was sent by the Council of Bremen to Cromwell; later he went with pupils to Oxford, where he stayed until 1657. As Boyle was the uncle of the pupils, Oldenburg early came in contact with this scientist, and devoted himself to experimental science. He combined interest in science with literary ability and unusual command of Latin, and he translated some of Boyle's and Steno's works into Latin (*Dictionary of National Biography*, s.v.; Birch, *History of the Royal Society of London*, III, 352).

[53] *Record of the Royal Society* (1901), p. 40.

[54] *Ibid.*, p. 41.   [55] *Ibid.*, p. 42.   [56] Weld, *op. cit.*, I, 128.

[57] *Ibid.*, p. 130. It must be mentioned that the King assigned to the society certain claims in Irish lands on which, however, it proved impossible to realize.

THE ROYAL SOCIETY

continuous pecuniary struggle which more than once threatened its very existence.[58] And yet the fact of its incorporation was a great step in advance, as the royal sanction in the eyes of the world transformed the workers from the type popularly known as "cranks" into respected investigators of new lines of truth.

The council proceeded immediately to the appointment of ninety-six Fellows of whom the following are the best known: Isaac Barrow, Theodore Hauk, Robert Hooke, Bishop Sprat, John Wallis, Willoughby, John Winthrop, and Christopher Wren.[59]

The statutes of the society were drawn up more than a year later. Within this time fell the important appointment of Robert Hooke as curator of the society. He had as a boy been prevented from study by headaches, and, left to himself, had sought diversion in mechanical toys. In 1653 he entered Oxford where his mechanical skill attracted wide notice, for example, he

[58] Weld, *op. cit.*, I, 128.

[59] *Record of the Royal Society* (1901), pp. 227 ff. The following table shows the number of new members added to the society in the succeeding years:

1663—33	1676— 9	1689— 4
1664—30	1677—12	1690— 1
1665—14	1678—12	1691— 5
1666—16	1679—12	1692—12
1667—33	1680— 6	1693— 8
1668—24	1681—26	1694— 3
1669— 5	1682— 9	1695— 9
1670— 4	1683— 8	1696—18
1671— 5	1684— 9	1697—12
1672— 4	1685— 8	1698—23
1673—14	1686— 5	1699— 3
1674— 6	1687— 4	1700— 7
1675— 5	1688— 7	

The dates of admission of the more prominent members added from time to time are the following, according to the *Record of the Royal Society* (1901): 1663, Huygens, Bathurst, Sir William Petty, Willis; 1664, Hevelius, Glanvill, Sam Pepys, Prince Rupert; 1666, Adrian Auzout; 1667, Richard Lower, John Collins, Ray; 1668, Malpighi; 1671, Grew, Newton; 1672, Cassini; 1673, Leibniz; 1676, Flamsteed; 1678, Halley; 1679, Leeuwenhoek; 1680, Papin; 1691, Count Marsigli; 1696, Viviani; 1698, Geoffrey; 1699, Menkenius; 1700, Sydenham.

showed the "artifices of flying" to John Wilkins. He studied astronomy with Seth Ward, and assisted Willis in chemistry, who in turn recommended him to Boyle. Hooke instructed Boyle in Euclid, and it was an act of generosity on Boyle's part to relinquish Hooke's services to the society. His duty as curator was "to furnish the society every day they met with three or four considerable experiments, expecting no recompense till the society get a stock enabling them to give it." The statutes which the society adopted in 1663 seem, in great part, his work. A few extracts from them[60] will give a clear idea of the society's aim and methods:

> The business of the society in ordinary meetings shall be to order, take account, consider and discourse of philosophical experiments and observations; to read, hear and discourse upon letters, reports and other papers containing philosophical matters; as also to view and discourse upon rarities of nature and art; and thereupon to consider what may be deduced from them or any of them, and how far they or any of them may be improved for use or discovery.[61]
>
> Two or more curators shall be appointed for every experiment or natural observation, that cannot conveniently be performed in the presence of the society, which curators .... shall .... meet together at time and place .... agreed .... to make the experiment or observation; and also shall jointly draw up a report of the matter in every such experiment and observation. ....[62]
>
> In all reports of experiments to be brought to the society, the matter of fact shall be barely stated without any prefaces, apologies, and rhetorical flourishes; and entered so in the register book by order of the society, and if any Fellow shall think fit to suggest any conjecture concerning the causes of the phenomena in such experiment, the same shall be done apart; and so entered into the register book.[63]

[60] Weld, *op. cit.*, II, 524 ff. (to be found in full).

[61] *Ibid.*, IV, Sec. V.   [62] *Ibid.*, V, Sec. III.

[63] *Ibid.*, Sec. IV. Cf. with the fuller statement quoted in Weld, I, 146 ff., taken from the manuscript volume of Hooke's papers, written in 1663:

"The business and design of the Royal Society is—

"To improve the knowledge of naturall things, and all useful Arts, Manufactures, Mechanick practices, Engynes and Inventions by Experiments—(not

THE ROYAL SOCIETY

The secretary shall have charge of the charter-book, statute-book, journal-book, letter-book[64] and register-book, "wherein shall be fairly written all such observations, histories and discourses of natural and artificial things—as also all such philosophical experiments, together with particular accounts of their processes as shall be ordered to be entered therein."[65] The office of the curator[66] was thus established:

The employment and business of the curator shall be to take care of the managing of all experiments and observations appointed—examining of

---

meddling with Divinity, Metaphysics, Moralls, Politicks, Grammar, Rhetorick, or Logick).

"To attempt the recovering of such allowable arts and inventions as are lost.

"To examine all systems, theories, principles, hypotheses, elements, histories, and experiments of things naturall, mathematicall, and mechanicall, invented, recorded or practiced, by any considerable author ancient or modern. In order to the compiling of a complete system of solid philosophy for explicating all phenomena produced by nature or art, and recording a rationall account of the causes of things.

"In the mean time this Society will not own any hypothesis, system, or doctrine of the principles of naturall philosophy, proposed or mentioned by any philosopher ancient or modern, nor the explication of any phenomena whose recourse must be had to originall causes (as not being explicable by heat, cold, weight, figure, and the like, as effects produced thereby): nor dogmatically define, nor fix axioms of scientificall things, but will question and canvass all opinions, adopting nor adhering to none, till by mature debate and clear arguments, chiefly such as are deduced from legitimate experiments, the truth of such experiments be demonstrated invincibly.

"And till there be a sufficient collection made of experiments, histories, and observations, there are no debates to be held at the weekly meetings of the Society, concerning any hypothesis or principal of philosophy, nor any discourses made for explicating any phenomena, except by speciall appointment of the Society or allowance of the President. But the time of the assembly is to be employed in proposing and making experiments, discoursing of the truth, manner, grounds and use thereof, reading and discoursing upon letters, reports and other papers concerning philosophicall and mechanicall matters, viewing and discoursing of curiousities of nature and art, and doing such other things as the Council or the President shall appoint."

[64] Statutes, c. 10, Sec. I.   [65] *Ibid.*, c. 11, Sec. I.

[66] The office of curator was given to Robert Hooke, first temporarily, then "for perpetuity," with a salary of thirty pounds, and apartments in Gresham College assigned as his residence.

science, arts, inventions; bringing in histories of natural and artificial things. Every .... curator .... shall be .... knowing in philosophy and mathematical learning, addicted to and well versed in observations, inquiries and experiments concerning natural and artificial things. Every .... curator .... shall be elected first for one year on probation and at the end .... for perpetuality.[67]

Offices of clerk, printer, and of operators of the society were also established.[68]

And now that we have seen how the society was established with its charter and statutes, let us commence to study somewhat closely those features whereby it helped to advance the progress of experimental studies during the first decades of its corporate existence.[69]

First, as to the personnel of the Fellows: There were fourteen noblemen, barons, and knights; eighteen esquires; eighteen physicians; five doctors of divinity; two bishops "for prevention of those panick, causeless terrors";[70] thirty-eight other members.[71] We have here an association not of scholars and learned men pre-eminently, but of amateurs interested in experi-

---

[67] Statutes, c. 11, Secs. II, III, and VIII.

[68] Clark, *op. cit.*, I, 290. It is of some interest that the position of operator was for years filled by Peter Stahl, the instructor of chemistry of the "Invisible College."

[69] The society held its meetings at Gresham College regularly until interrupted (from June, 1665, to February, 1666), first by the Great Plague, during which period they met at Boyle's in Oxford occasionally; and again in September, 1666, by the Great Fire. After this latter catastrophe the quarters at Gresham College were needed to accommodate citizens. Hence the society met (1666–73) at Lord Howard's Arundel House. It is a sign of their prominence that in 1673 a committee of aldermen, magistrates, and Gresham professors waited on them asking them to return to Gresham College. The invitation was accepted, in view of the fact that the curator's quarters were there and all apparatus on hand (Weld, *op. cit.*, I, 182 ff., 198).

[70] Glanvill, *op. cit.*, Preface.

[71] *Record of the Royal Society* (1901), p. 228.

mental science, a fact in which Sprat revels.[72] Indeed, appeal was specially made to the amateur, who could not give much of his time.

They [the Royal Society] require not the whole time of any of their members, except only of their curators. From the rest they expect no more but what their business, nay, even their very recreation can spare. It is continuance and perpetuity of such philosophical labors to which they trust.[73]

The appeal is specially made to the man without university learning. Sprat wrote:

It suffices if many be plain, diligent and laborious observers who though they bring not much knowledge, yet bring their hands and eyes incorrupted. . . . . Men did generally think no man was fit to meddle in matters of consequence but he that had bred himself up in a long course of discipline for that purpose. . . . . Experience . . . . tells us greater things are produced by free way than the formal.[74]

Hooke emphasized that it was much to the advantage of the Royal Society that many merchants and business men were joining its ranks,[75] also because thereby the practical advantage of financial help might accrue.

Turning to a still more essential point—how the society conducted its investigations—we find that they did not have as yet a laboratory in the proper sense of the word. The curator's (Robert Hooke's) rooms in Gresham College were the nearest

[72] Sprat, *op. cit.*, p. 71: "We find noble rarities to be every day given in, not only by the hands of the learned, but from shops of mechanics, voyagers of merchants, ploughs of husbandmen, gardens of gentlemen."

[73] *Ibid.*, p. 333. Sprat exults that every class of men has joined (*ibid.*, p. 63): "They have freely admitted men of different religions, countries and professions of life." *Ibid.*, p. 130: "Physicians contribute, though they have had their own college 150 years; statesmen, soldiers and churchmen."

[74] *Ibid.*, pp. 72 f.

[75] Hooke, *op. cit.*, Preface. "Many of their number are men of traffic which is a good omen that their attempts will bring philosophy from words to action, seeing that men of business have had such great share in the first foundation. . . . . Several merchants, men who act in earnest, have adventured considerable sums of money to put in practice what some of our members have contrived."

approach to it. The creation of a laboratory and designing of apparatus were of course the first care of the society.[76] But there were great difficulties in the way, as is evident from Hooke's letter to Boyle:

> We are now undertaking several good things, as the collecting a repository, the setting up a chemical laboratory, a mechanical operatory, an astronomical observatory, and an optick chamber; but the paucity of the undertakers is such, that it must needs stick, unless more come in, and put their shoulders to the work. We know, Sir, you can and will do much to advance these attempts.[77]

And so insufficient was the money to purchase such philosophical apparatus as was required that the society was frequently under the necessity of soliciting Boyle to lend his personal apparatus for experiments.

Sprat[78] gives an enumeration of the society's instruments—"those new ones which they themselves or some of their members have invented or advanced":

> Instrument for finding a second of time by the sun.
> Three several quadrants. New instrument for taking angles by reflection.
> New kind of back-staff for taking altitude of the sun.
> Hoop of all fixed stars in the Zodiac.
> Copernican sphere.
> Instruments for keeping time exactly . . . . with pendulums and without.

---

[76] Weld, *op. cit.*, I, 185. Hooke wrote to Boyle in 1666: ". . . . I design very speedily to make an operatory, which I design to furnish with instruments and engines of all kinds, for making examination of the nature of bodies optical, chemical, mechanical. . . . . I am now making a collection of natural rarities."
I quote two instances I find of orders for and description of apparatus: "The Operator is called upon to have ready cylinders of mercury for observing the ascent and descent thereof, according to various constitutions of air" (Birch, *op. cit.*, I, 304). And again, Sir Robert Moray described a kind of furnace which puts air into a flame by pipes, being outwardly heated and blown by bellows; by which means some persons pretend to melt ore into water" (*ibid.*, p. 144).

[77] Weld, *op. cit.*, I, 186.

[78] Sprat, *op. cit.*, pp. 246 ff.

A universal standard or measure of magnitudes by help of pendulum.
New pendulum clock.
Pendulum clock showing equation of time.
Three new ways of pendulums for clocks.
Pendulum watches.
Instruments for compressing and rarefying air; wheel barometer.
New kind of scales to examine gravity; to see whether the attraction of the earth be no greater in some parts of the earth than in others.
Exact scales for magnetical experiments.
Several accurate beams.
Magnetical instruments.
New kinds of levels.
New Augar for boring.
New instrument for fetching any substance from bottom of sea.
New bucket.
Two new ways of sounding the depths of the sea.
Instruments for finding velocity of swimming bodies.
Instrument of great height with glass windows, to be filled with water for examining velocity of bodies . . . . by their descent.
Instrument for measuring and dividing time of their descent, serving for examining swiftness of bodies descending.
Bell for diving.
Instruments for diver.
New spectacles for diver.
New way of conveying water to diver.
Instrument for measuring swiftness of wind and for raising continual stream of water.
Several thermometers: one for examining all degrees of heat in flames and fires made of several substances.
New standard for cold.
Instrument for planting corn.
Four sorts of hygroscopes.
Several kinds of ways to examine water.
Several engines for finding and determining the force of gun powder.
Instrument for receiving and preserving the force of gun powder so as to make it applicable for performing of any motion desired.
Several instruments for examining recoiling . . . . of guns.
Instruments to improve hearing.
Models of chariots.

Chariot-way-wiser, measuring length of way of chariot.[79]
Instrument for making screws.
Way of preserving impression of seal.
Instrument for grinding optic glasses; double telescope, several telescopes of divers length. [Sprat here remarks that the English have great advantage of late years through their art of making glass finer and more serviceable for microscopes and telescopes than that of Venice.]

Grew, in his book *Musaeum Regalis Societatis; or a Catalogue and Description of the Natural and Artificial Rarities belonging to the Royal Society and preserved at Gresham College* . . . ., published in 1681, enumerated among the society's instruments in that year the following:

Air pump (Boyle); condensing engine (one of brass and of glass); weather clock (Wren and Hooke); instrument to measure quantity of rain (Wren); lamp furnace (Hooke); semi-cylindric lamp (Hooke); model of eye, burning glass, hollow glass (Wilkins); one microscope that magnifies one hundred diameters, lesser microscope, three instruments to help hearing (Wilkins); hydrostatic scales, anatomic instruments, reflecting telescope (Newton); arithmetic instrument (Hooke); way-wiser (Wilkins); model of a two-bottomed ship (Sir W. Petty); dipping needles, canoe, guns.[80]

Valuable instruments were presented to the society by their inventors as, for example, Newton's telescope, Huygens' lens, and Papin's boiler.

Closely connected with the question of the laboratory was that of the acquisition of a collection of rarities, so important to the seventeenth century. The society started such a rarity cabinet with the gift of Daniel Colwall, and Grew gives a detailed account[81] of the objects in the possession of the society, which would correspond to the anatomical, zoölogical, and botanical collections of today. These were considerable for the time, but were comparatively on a much lower level than the instruments. The "Musaeum" contained, indeed, many a queer

[79] Such a device is described by Vitruvius.
[80] Nehemiah Grew, *op. cit.*, Part IV, pp. 351 ff.
[81] *Op. cit.*

THE ROYAL SOCIETY

specimen, if we are to trust to the comment of as learned a man as Magalotti:

> Amongst these curiosities, the most remarkable are: an ostrich whose young were always born alive; an herb which grew in the stomach of a thrush; and the skin of a moor, tanned, with the beard and hair white; but more worthy of observation than all the rest, is a clock, whose movements are derived from the vicinity of a loadstone, and it is so adjusted as to discover the distance of countries at sea by the longitude.[82]

Having thus examined the "inventory" of the Royal Society, let us turn to their work itself.

As to its organization, we find an interesting account by Sprat which shows that much actual experimenting was done at the homes of the Fellows, and which throws light on the allotment of topics and preliminary preparations before the stage of experimentation was reached.[83] The society as a whole discussed "without rigor or criticism" the problem on hand, and what could be found about it in books;[84] then one would perform the experiment privately and bring "the history of the trial to the society."[85] Experimental work along physiological and anatomical lines was done in the private laboratories of members—dis-

---

[82] Weld, *op. cit.*, I, 219.

[83] "Men who selected their own topics were allowed to experiment privately, and expenses were allowed from the common stock; or the society made distribution, and deputed whom it saw fit for the experiment; they either allotted the same work to several men, or else joined them into a committee" (Sprat, *op. cit.*, p. 84); ". . . . The usual course was, when they appointed the trial, to propose one week some particular experiment to be prosecuted next and to debate beforehand concerning all things that might conduce to better carrying on" (*ibid.*, p. 95). Another glimpse of the society at work is given in Birch, *op. cit.*, III, 346: "November 1st, 1677, there were produced a great many exceedingly small and thin pipes of glass of various sizes, some ten times, others a tenth as big as a hair, to test Leeuwenhoek's experiments." When micro-organisms were not found, it was decided to get a better microscope. November 8, 1677, the better microscope was obtained, yet the animals in the pepper water were not seen. December 6, 1677, they were seen.

[84] Sprat, *op. cit.*, p. 95: "The best experimenters are least versed in books."

[85] *Ibid.*, p. 97.

sections, perhaps, at the Royal College of Physicians. At every meeting the curator, Robert Hooke, exhibited experiments. In addition to the intrinsic interest of what was shown, it was designed, according to Thomson,[86] to put the spectators on the right way of investigation and to teach them how to draw legitimate conclusions from their own observations. But he ventures the criticism that "this seems not a good way, as few discoveries were made during exhibitions, even when the experimenters were successful in private labors."

The work itself was so wide in scope that it defies classification. Sprat (1667) naïvely divided it into eleven types: work dealing with (1) fire; (2) atmosphere, sound, breathing, experiments with rarefying engine; (3) water; (4) metals and stones; (5) vegetables; (6) medical matters; (7) qualities sensible as to heat; (8) other qualities, rarities; (9) light, sound, color; (10) motion; (11) matters chemical, mechanical, optical.[87] But as a matter of fact most experiments, new and old, that could be done were performed by the society,[88] particularly under the guidance of the indefatigable Hooke, equally efficient in the most varied branches of investigation. To give a hint of the significance of the early work of the Royal Society, I shall mention a few of the most famous lines of experimentation.

As in the case of the Cimento, much effort was directed to the invention and improvement of instruments. For instance, Hooke improved the telescope, devised a spring for watches, etc.[89] The study of micro-organisms was a topic especially cul-

[86] Thomson, *History of the Royal Society*, I, 6.

[87] Sprat, *op. cit.*, pp. 215 ff.

[88] "Eight committees were appointed on March 30, 1664: 1, Mechanical, consisting of 69 members; 2, Astronomical and Optical, consisting of 15 members; 3, Anatomical, consisting of Boyle, Hooke, Dr. Wilkins, and all physicians of the society, and seven other Fellows; 5, Geological, consisting of 32 members; 6, for histories of trades, consisting of 35 members; 7, for collecting all the phenomena of nature hitherto observed, and all experiments made and recorded, consisting of 21 members" (Weld, *op. cit.*, I, 174 ff.).

[89] About this Hooke had violent contests with Huygens.

tivated by the society, as it was of particular interest to its curator, Robert Hooke. He had improved the microscope, and with this showed the society the appearance of a common moth, of crystals, molds, seeds, gnats, scales, and fleas.[90] So delighted was the company that "Mr. Hooke was charged to bring one microscopical observation at every meeting."[91]

The study of micro-organisms suggested the question of spontaneous generation. We hear of Boyle's experiments along these lines, and again upon "Sir Robert Moray's inquiring how worms could come from vinegar, Dr. Goddard suggested that vinegar probably had crude waterish matter, which, turning into slimy substance, might breed worms."[92]

Problems of dynamics were investigated under the direction of Hooke; he devised an instrument to demonstrate Galileo's laws of falling bodies; made extensive studies of the laws of free fall from St. Paul's dome, and in the course of these studies claimed to have discovered the law that attraction varies inversely as the square of distances.[93] Problems of optics, especially of the nature of color and colors of thin plates and soap bubbles, were studied by the tireless curator of the society.

Another series of experiments, wherein the society, under Boyle's and Hooke's leadership, did pioneer work, was the study of the relation of air to breathing and burning. Borelli had shown that not merely the movement of the chest but air was essential to life.[94] This Hooke demonstrated by putting dogs and vipers under the receptacle of a rarefying engine, asphyxiating and reviving them in turn.[95] But how utterly at sea as to the

[90] Birch, *op. cit.*, I, 216.

[91] *Ibid.*, p. 215. The collection of these observations forms one of the earliest books printed by the society, Hooke's *Micrographia*, a most interesting volume. Its Preface, often quoted by me, is a veritable hymn to inductive science and the Royal Society.

[92] Birch, *op. cit.*, II, 49.

[93] This involved him in a priority contest with Newton.

[94] Foster, *op. cit.*, p. 180.   [95] Birch, *op. cit.*, II, 188.

real issues all these experimenters were is shown by Hooke's inquiry as to whether air is consumed or generated in burning.[96] Boyle had shown that both candlelight and life were extinguished in a vacuum. At Dr. Ent's suggestion the society put animals and burning coal simultaneously into the receptacle of an air pump, and attempted to study the interrelation of burning and breathing.[97] These investigations directed the interest of the company toward questions of a physiological nature, and experiments of injecting liquors in veins,[98] etc., were made.

Great interest was aroused during 1666 and 1667 by experiments in the transfusion of blood.[99] On June 20, 1666, the first notice of a remarkable operation for the transfusion of blood was recorded in the journal-book. Dr. Wallis related the success of the experiment made at Oxford by Dr. Lower,

of transfusing the bloud of one animal into the body of another, viz., that having opened the jugular artery of a mastiff, and injected by the means of quils the bloud thereof into the jugular vein of a greyhound, and opened also a vein in the same greyhound, to let out so much of his bloud as was requisite for the receiving that of the mastiff; the mastiff at last died, having lost almost all his bloud, and the greyhound having his vessels closed, survived, and ran away well.[100]

This account created great interest among the Fellows, and the identical experiment was tried before the society, November 14, 1666. It was repeated by bleeding a sheep into a mastiff, and thus the possibility of blood transfusion between different animals was established.[101] A few months later an account was received from Paris of two experiments made at the Académie des Sciences, on a youth and an adult, whose veins were open and injected with the blood of lambs. The experiment, according to

[96] *Ibid.*, III, 77.
[97] *Ibid.*, II, 12.
[98] Weld, *op. cit.*, I, 273.
[99] These will be described somewhat fully on account of the interest this question meets with in recent research.
[100] Weld, *op. cit.*, I, 192.   [101] *Ibid.*, p. 195.

the account, succeeded so well that the Royal Society became anxious to repeat it, and Sir George Ent suggested that it would be most advisable to try it upon "some mad person in Bedlam."[102] But Dr. Allen, of the insane asylum, "scrupled to try the experiment." It was ultimately performed on a poor student, named Arthur Coga, who, being in want of money, offered himself to the investigators for a guinea. The operation was performed by Dr. Lower and Dr. King at Arundel House on November 23, 1667, in the presence of several spectators, among whom were the Bishop of Salisbury and some members of Parliament. Sheep blood was transfused through the patient's veins.[103] The experiment was repeated at a public meeting of the society on December 12 following, when eight ounces of blood were taken from Coga, and about fourteen ounces of sheep's blood injected.[104] While success attended this one experiment in England, in Paris after one fatal result the courts interdicted further transfusion of blood, and this line of research was likewise stopped in England.

Another most important feature of the laboratory work of the Royal Society was the repetition and testing of experiments done by other scientists and societies. The company tested and repeated the series of experiments of Boyle with the vacuum;[105] Huygens' work on the isochronous pendulum;[106] Mariotte's investigations relating to the blind spot in the eye;[107] the experi-

[102] *Ibid.*, p. 200.

[103] *Ibid.*, pp. 220 ff. Oldenburg, in a letter to Boyle, giving an account of the experiment, observes: "Dr. King performed the chief part of it with great dexterity, and so much ease to the patient, that he made not the least complaint, nor so much as any grimace during the whole time of the operation; that he found himself very well upon it, his pulse and appetite being better than before, his sleep good, his body as soluble as usual. . . . ." A person asking why the patient had not the blood of some other creature instead of that of a sheep transfused into him, he answered: "Sanguis ovis symbolicam quandam facultatem habet cum sanguine Christi, quia Christus est Agnus Dei."

[104] *Ibid.*, p. 222.   [106] *Ibid*, II, 489.
[105] Birch, *op. cit.*, IV, 259.   [107] *Ibid.*, p. 281.

ments of Steno, De Graaf, of Van Helmont,[108] and of Guericke.[109] When in 1667 the *Saggi* of the Cimento were presented to the society, the task of reviewing them was assigned to a special committee, and the experiments there described, especially those made with "Boyle's engine," were tested.[110]

It was characteristic of the type of men in the society that many of their investigations dealt with eminently practical subjects.[111] Sprat described this phase of their work as follows:

> They have propounded the composing a catalogue of all trades, works, and manufacture . . . . taking notice of all physical receipts or secrets, instruments, tools, and engines, manual operations or sleights. . . . . They recommended advancing the manufacture of tapestry, silk-making, melting of lead ore with pit coal . . . . making trials of English earths to see if they will [not do] for perfecting of the potter's art. They have [compared] soils and clays for making better bricks and tiles. They started the propagation of potatoes and experiments with tobacco oil.[112]

Investigations were made for bettering wines, improving methods of brewing ale and beer,[113] manuring with lime,[114] devising a new cider press, a lamp for hatching eggs.[115] They studied how foils were made in Germany,[116] and questions relating to the designing of carriages.[117] Sir Wm. Petty's ship aroused the greatest interest, and after it was wrecked Sprat felt much as Germany felt about the Zeppelin machine—it should be

[108] *Ibid.*, p. 239.
[109] *Ibid.*, IV, 259.
[110] *Ibid.*, II, 256 f.
[111] As Hooke (*op. cit.*, Preface) puts it: "[They] acknowledge their most useful information to arise from common things and diversify their most ordinary operations upon them."
[112] Sprat, *op. cit.*, pp. 190 ff.
[113] Birch, *op. cit.*, II, 489.
[114] *Ibid.*, I, 162.
[115] *Ibid.*, p. 246.   [116] *Ibid.*, II, 259.
[117] "Hauk is to bring a draft of great wagons of Lübeck and Hamburg carrying sixteen persons and nine horses three abreast" (*ibid.*, p. 28).

THE ROYAL SOCIETY

tried again at public expense.[118] Winthrop read a paper concerning the convenience of building ships in some of the northern parts of America, on account of the great store of good oak and pine and sawmills there.[119] As early as October 15, 1662, the King declared that no patent should be granted for any philosophical, mechanical invention until examined by the society.[120] Hence there were submitted to the Fellows many engines; for instance, one to make linen cloth, which is interesting as a forerunner of Hargreave's machine.[121] Even from Germany models of machinery, e.g., a printing press, were presented to them.[122]

Another phase of the work of the society—as we saw above—was the inquiry into conditions in foreign lands. We find their methods in these inquiries described in detail by Sprat,[123] who also gives a complete example of such queries and replies from Java.[124] When people started on voyages, it was customary to ask the society what inquiries the travelers should make.[125] So, systematically, the fabulous notions concerning foreign countries were made to give way to rational ideas about them.

[118] Sprat, *op. cit.*, p. 240.
[119] Birch, *op. cit.*, I, 112.   [120] Weld, *op. cit.*, I, 137.
[121] *Philosophical Transactions* (abridged), I, 501 f. It was described in the following manner: "The advantages of the engine are these—one mill will set ten or twelve of these looms at work. One boy will serve to tie the threads at several looms."
[122] Cantor, *op. cit.*, III, 36.
[123] "First they employ Fellows to examine treatises, etc., of countries; they employ others to discourse with seamen, travellers, tradesmen and merchants; then they compose a body of questions about observable things. Then the Fellows would start correspondence with the East Indies, China, St. Helena, Teneriff, Barbary, Morocco . . . ." (Sprat, *op. cit.*, p. 155). "In this our chief and most wealthy merchants and citizens many have assisted with their presence and contributed their labors and helped correspondence; employed factors abroad to answer inquiries; they have laid out in all countries for observations and gifts . . . ." (*ibid.*, p. 129).
[124] *Ibid.*, pp. 158 ff.
[125] Birch, *op. cit.*, I, 297, quotes an instance of such inquiry made by a German prince traveling in Egypt.

The important question as to whether or not the Royal Society should undertake educational work early presented itself. Forming a college—as they were permitted to do by their charter—seemed advisable, as it would, in Oldenburg's words,

in all likelihood establish our institution and fix us who are now looked upon but as wanderers and using precariously the lodgings of other men in a certain place, where we may meet, prepare and make our experiments and observations, lodge our curators and operators, have our laboratory, observatory and operatory, all together.[126]

Then it was thought that a college might serve as a source of much-needed income.[127] The majority, however, under the lead of Boyle, were averse to turning the society's work of investigation into teaching channels.[128]

Another phase of educational work, that of giving lectures—today so important a part of the function of learned societies—was initiated in June, 1664, by Sir John Cutler, who founded a professorship of mechanics, and, with the concurrence of the Council of the Royal Society, settled an annual stipend of fifty pounds during life on Hooke, empowering the President, Council, and Fellows of the society to appoint the subjects and number of lectures.[129] The only other endowed lectureship in these early years was due to a bequest by Dr. Croone's widow.[130]

---

[126] Weld, *op. cit.*, I, 208.

[127] Christopher Wren had plans for such a college all drawn, which sound most attractive and modern. Weld gives these plans in full: "It contains in the foundations, first a cellar and a fair laboratory; then a little shop or two, for forges and hammer works, with a kitchen and little larder. . . . . In the third story are two chambers with closets, for the Curators . . . . [there is] a little passage-gallery the whole length of the building, for trial of all glasses and other experiments that require length. On one side of the gallery are little shops all along for operators. . . . . The platform of lead is for traversing the tubes and instruments, and many experiments. In the middle rises a cupola for observations and may be fitted, likewise, for an anatomy theatre; and the floors may be so ordered, that from the top into the cellar may be made all experiments for light" (*ibid.*, pp. 212 f.).

[128] *Ibid.*, p. 214.   [129] *Ibid.*, p. 173.

[130] "One-fifth of the clear rent of the King's Head tavern be vested in the Royal Society, for the support of a lecture and illustrative experiment, for the

THE ROYAL SOCIETY

Later (1673) the society conceived the idea of public meetings, apparently every Thursday, both as a source of education and of income.[131] The character of the public lectures may be inferred from a program arranged for the entertainment of the King.[132]

But we must turn now to another feature of the society's work, scarcely less important than that done in the laboratory, the establishment of intercommunication between scientists. However extensive the duties of the curator seemed, they were slight compared to the labor that rested on the shoulders of the second secretary, Oldenburg. He describes his duties as follows:

> Attends meetings. . . . . Noteth the observables said and done there . . . . digesteth them in private . . . . takes care to have them entered in Journal and Register book . . . . reads over and corrects all entries, solicits performance of tasks recommended and undertaken . . . . writes all letters abroad and answers returns made to him, entertaining correspondence with at least thirty persons, employs a great deal of time and takes much pain in satisfying forrain demands about philosophical matters, disperses far and near store of directions, inquires for the societies purpose: Q. Whether such a person ought to be left unassisted.[133]

The extensive correspondence Oldenburg kept up was collected in the voluminous folios of the letter-books. It was invaluable as it formed the means of keeping in touch with what was

---

advancement of natural knowledge on local motion, or [conditionally] on such other subject, as in the opinion of the President for the time being, should be most useful in promoting the objects for which the Royal Society was instituted" (*ibid.*, p. 289).

[131] *Ibid.*, p. 247. They passed a resolution rendering it imperative on every member of the existing council "to provide an experimentall discourse for the society, to be made at some publique meeting within the year, either by himself, or by some other member of the society, or to pay forty shillings."

[132] Boyle should show experiments on cohesion, air pressure, change of color in chemicals, heating magnet; Goddard should experiment with the hygroscope; Dr. Ent should show the anatomy of the lobster and oyster; Clarke, his collection of insects; Hooke should show experiments with a thermometer, an artificial eye, and casting of pictures on the wall (Birch, *op. cit.*, I, 312).

[133] *Dictionary of National Biography*, s.v. "Oldenburg."

done for science in other countries. First, there were the letters with the regular foreign correspondents, for example, Hevelius, the Dantzig astronomer; Huygens; Malpighi,[134] sought eagerly both by Oldenburg and the foreigners.[135] Then scientists from everywhere communicated their ideas, discoveries, and observations to Oldenburg;[136] indeed, the reading of these foreign letters formed an essential feature of the sessions of the society. Through this correspondence the Royal Society was in constant touch with other nations, so that at times it seems an international rather than an English body; every important experiment, every important article, was communicated to it almost as soon as it was published. Hence when a foreign savant arrived at London he found the society fully instructed in his work. How important this element of correspondence was is shown by the fact that after the death of Oldenburg, anxious inquiries were made about its continuance.[137]

Besides letter-writing, the industrious secretary had to follow all scientific articles printed in contemporaneous publications. This was fortunately for Oldenburg a slight task, at the commencement of his secretaryship, but in succeeding years it involved the reading of articles in the *Journal des Sçavans, Miscellanea curiosa,* and other scientific journals.[138] This task Oldenburg conscientiously performed, and reported on important articles at the sessions.[139]

Out of Oldenburg's wide correspondence and his record of the experiments of the society developed his publication of a regular periodical scientific paper, *The Philosophic Transactions.* The *Journal des Sçavans,* antedating it by three months, deprives

[134] Birch, *op. cit.,* II, 333.

[135] Leopold of Medici, cardinal in 1668, obtained the Pope's permission to correspond with the Royal Society (*ibid.,* p. 335).

[136] To give instances: "Letter received from Steno; he had found method of working certain lenses" (*ibid.,* p. 100); "Governor Winthrop from Massachusetts reports to Oldenburg on [unsuccessful] experiments" (*ibid.,* I, 280).

[137] *Ibid.,* III, 418.   [138] See below, chap. vii.   [139] Birch, *op. cit.,* II, 358.

it of the honor of being the first scientific periodical, but it was the first ever published under the auspices of a society which was destined to last to the present time. The Royal Society had early planned a regular publication;[140] but as nothing was done Oldenburg decided, as a private venture, to publish monthly the matters of most importance which the members of the society or foreign scientists communicated to him. This plan was accepted by the society. On the first of March, 1664–65, it was ordered at a meeting of the Council

> that the *Philosophical Transactions,* to be composed by Mr. Oldenburg, be printed the first Monday of every month, if he have sufficient matter for it; and that the tract be licensed by the Council of the society, being first reviewed by some of the members of the same; and that the President be now desired to license the first papers thereof.[141]

In conformity with this order, the first number of the *Transactions* appeared on Monday, March 6.[141]

Thus originated the famous series of volumes[142] to which Huxley pays the following tribute:

[140] This is evident from the subjoined notice of Robert Hooke (1663): "They conceive many useful and excellent observations may be collected into a general repository, where inquisitive men be sure to find them safely and carefully preserved. . . . . They resolve to gratify all that communicate, with suitable returns of such experiments, observations, and inventions of their own, or advertisements from others of their correspondents. And that you may understand what parts of naturall knowledge they are most inquisitive for at this present, they designe to print a Paper of advertisements once every week or fortnight at furthest, wherein will be contained the heads or substance of the inquiries they are most solicitous about, together with the progress they have made and the information they have received from other hands, together with a short account of such other philosophicall matters as accidentally occur, and a brief discourse of what is new and considerable in their letters from all parts of the world, and what the learned and inquisitive are doing or have done in physick, mathematicks, mechanicks, opticks, astronomy, medicine, chymistry, anatomy, both abroad and at home" (Weld, *op. cit.,* I, 148 f.).

[141] *Ibid.,* p. 177.

[142] Upon Oldenburg's death, Dr. Nehemiah Grew, the succeeding secretary, published five contributions within two years. Then the publication was intermitted for three years, and Dr. Hooke published the *Philosophical Collections*

If all the books in the world except the *Philosophical Transactions* were destroyed, it is safe to say that the foundations of physical science would remain unshaken, and that the vast intellectual progress of the last two centuries would be largely, though uncompletely, recorded.[143]

Oldenburg's dedication to the Royal Society clearly shows that the publication was his work and not the society's.[144] His Introduction is a clear statement of the purposes of scientific publications:

> Whereas there is nothing more necessary for promoting the improvement of philosophical matters, than the communicating to such as apply their studies and endeavours that way, such things as are discovered and put in practice by others; it is therefore thought fit to employ the press, as the most proper way to gratifie those whose engagement in such studies and delight in the advancement of learning and profitable discoveries doth entitle them to the knowledge of what this kingdom, or other parts of the world do afford, as well as of the progress of the studies, labours and attempts of the curious and learned in things of this kind, as of their compleat discoveries and performances; to the end that such productions being clearly and truly communicated, desires after solid and usefull knowledge may be further entertained, ingenious endeavours and undertakings cherished, and those addicted to or conversant in such matters may be invited

―――――――――

("Accounts of Physical, Anatomical, Chymical, Mechanical, Astronomical, Optical, and other mathematical and philosophical experiments and Observations") which have always been considered a portion of the *Transactions* (1681–82); then for a year the publication was discontinued, owing to the limited sale and small profit to the secretary; in 1683 Dr. Robert Plot revived the old *Philosophical Transactions,* and he, Musgrave, and Dr. Halley published them continuously, but not monthly, from 1683 to 1687. They were revived again in 1691, and from thenceforth have been published uninterruptedly (Thomson, *op. cit.,* p. 7).

[143] Thomas H. Huxley, *On the Advisableness of Improving Natural Knowledge.*

[144] "To the Royal Society. . . . . In these rude collections, which are only the gleanings of my private diversions in broken hours, it may appear that many minds and hands are in many places industriously employed, under your countenance, and by your example, in the pursuit of those excellent ends, which belong to your heroical undertakings. . . . . And thus have I made the best use of some of [my minutes of leisure] . . . . to spread abroad encouragements, inquiries, directions and patterns, that may animate and draw on universal assistance" (Weld, *op. cit.,* I, 178).

and encouraged to search, try and find out new things, impart their knowledge to one another, and contribute what they can to the grand design of improving natural knowledge, and perfecting all philosophical arts and sciences. . . . .

The range of articles can be seen from the contents of the first number:

An Accompt of the improvement of Optick Glasses at Rome. Of the Observation made in England of a Spot in one of the Belts of the Planet Jupiter. Of the Motion of the late Comet predicted. The heads of many new Observations and Experiments, in order to an Experimental History of Cold, together with some thermometrical discourses and experiments. A relation of a very odd monstrous Calf. Of a peculiar Lead Ore in Germany, very useful for essays. Of an Hungarian Bolus, of the same effect with the Bolus Armenus. Of the new American Whale-fishing about the Bermudas. A Narrative concerning the success of the pendulum watches at sea for the longitudes; and the grant of a Patent thereupon. A Catalogue of the Philosophicall Books publisht by Monsieur de Fermat, Counsellour at Tholouse lately dead.

It is characteristic that three of these articles deal with scientific instruments. Indeed, matters relating to the grinding of lenses, the improvement of the telescope, the construction and use of barometers, contesting opinions of instrument-makers, occur continually in the pages of the publication.

The next conspicuous feature of the *Transactions* was that they constituted an international battle ground of scientific opinions; for example, when Auzout had a different view from Cassini, they communicated, through the London publications. Reviews of foreign books, extracts from the *Journal des Sçavans*, reports of the proceedings of the Académie des Sciences, were fairly permanent features. The writings of Boyle and reviews of his works fill many a folio page, clearly showing that he was the conspicuous figure in science in England before Newton became prominent. Discussions and reports of astronomical and physiological works are very numerous. The greatest number of papers are on experimental physics and biological sciences.[145] Indeed the Preface, with which the society's publications were renewed in 1683, sums up best their most essential feature:

> They are a specimen of many things which lie before them [i.e., the society], contain a great variety of useful matter, are a convenient register for the bringing in and preserving many experiments which, not enough for a book, would else be lost, and have proved a very good ferment for the setting of men of uncommon thoughts in all parts awork.

It is undoubtedly due to the sturdy perseverance of Oldenburg that this enterprise was launched amid forbidding financial difficulties. During his life the one hundred and thirty-six monthly publications (1665–77), filling the first twelve volumes, came out regularly, though they never yielded him a greater profit than four pounds a year, generally less.[146]

The great interest the publication aroused is shown by the fact that the *Transactions* were published in Latin in Frankfurt in 1671; in Leipzig, 1674; and also in Amsterdam, 1671.[147]

A most important phase of the society's work is yet to be touched upon, the publication of scientific works, by both mem-

---

[145] It is impossible to give an account of the contents of the volumes of the *Philosophical Transactions*, but some notion of the relative attention given by the society to the various subjects from 1665 to 1683 can be obtained from the proportion of papers devoted to each as grouped in the Index of the *Transactions* (abridged by Charles Hutton, George Shaw, Richard Pearson, 1809): "Natural Philosophy, Acoustics, Astronomy, Hydraulics, Magnetism, Meteorology, Optics, Pneumatics, 263; Agriculture, Antiquities, Voyages, 41; Anatomy, Physiology, Surgery, Medicine, Pharmacy, Chemistry, 131; Natural History, Botany, Mineralogy, Zoölogy, 237; Chronology, Geography, Mathematics, Mechanics, Navigation, 47; discussion of books, 395."

[146] Weld, *op. cit.*, I, 260. In a MS letter of Oldenburg's to Boyle, dated London, December 19, 1665, we have the following account of the sale of the *Transactions*: "Mr. Davis [the printer] tells me that of the first *Transactions* he printed he had not vended above three hundred, and that he fears there will hardly sell so many as to repay the charge of paper and printing, so that it seems my pains and trouble would be of no avayle to me" (*ibid.*, p. 182).

[147] *Academia Caesarea Leopoldina Carolina Germanica Naturae Curiosorum Miscellanea Curiosa sive Ephemeridum Medico Physicorum Germanorum* (referred to as *Miscellanea Curiosa*), Vol. XXV*, Index.

bers and foreign scientists. The first book they published was Evelyn's *Sylvia*,[148] soon afterward Hooke's *Micrographia*.[149] The society has the distinction of having helped John Ray[150] in 1667 to publish the work of his friend and patron, Sir Francis Willoughby, *History of Fishes*, a notable work in zoölogy.[151] It also arranged for the translation of the *Saggi* of the Accademia del Cimento,[152] and for the translation of the valuable *History of Animals* which the Académie des Sciences had issued. Similarly, works of foreign scientists were published.[153]

The notice of scientific works through correspondence, and their publication both in the *Philosophical Transactions* and in book form, was a great encouragement to scientists, as is best shown by the relation of the society to Leeuwenhoek. The first notice of Leeuwenhoek in the records of the Royal Society occurs in a letter in which Graaf wrote to Oldenburg (1673) that "one Mr. Leeuwenhoek hath lately contrived microscopes, excelling those that have been hitherto made." A short communi-

[148] Thomson, *op. cit.*, I, 65.

[149] Hooke, *op. cit.*, Preface. He states: "All was undertaken in prosecution of the design of the Royal Society."

[150] John Ray was among the society's most famous members; he and Willoughby traveled extensively in England and on the Continent for the purpose of collecting plants and animals and systematizing them. Ray published these results as *Historia plantarum* in three volumes, at the expense of Willoughby's widow, but later works on snakes and insects were published at the society's expense (Carus, *op. cit.*, p. 430).

[151] Weld, *op. cit.*, I, 310. This publication cost the society four hundred pounds and it exhausted the treasury to such an extent that the salaries even of their officers were in arrears, and Halley was paid fifty pounds as "fifty books on fishes."

[152] Waller, *op. cit.*

[153] Weld, *op. cit.*, p. 227. It is deserving of record that the celebrated Malpighi, while holding the professorship of medicine at Messina, sent his work, *Dissertatio epistolica de Bombyee*, to the society, with a request that it might be issued under their auspices. Moreover, an evidence of the high esteem in which the judgment of the Royal Society was held was shown by the fact that foreign scientists, such as De Graaf, Swammerdam, and Leibniz dedicated their books to the society.

cation from Leeuwenhoek accompanied the letter in which he described the structure of a bee and a louse. From this period Leeuwenhoek was in the habit of constantly transmitting to the society all his microscopical observations and discoveries, and three hundred and seventy-five papers and letters of Leeuwenhoek, extending over a period of fifty years, are preserved in the archives of the society.[154] He published in the *Philosophical Transactions* his paper on *Red Blood Corpuscles,* and in 1676 his letter on a great number of observations concerning *animalculae.* The gratitude he felt toward the society for the help they gave him will be seen by the following extract from one of his letters:

> I have a small black cabinet, lacker'd and gilded, which has five little drawers in it, wherein are contained thirteen long and square tin boxes, covered with black leather. In each of these boxes are two ground microscopes, in all six and twenty; which I did grind myself and set in silver; and an account of each glass goes along with them.
>
> This cabinet, with the aforesaid microscopes (which I shall make use of as long as I live) I have directed my only daughter to send to your Honors, as soon as I am dead, as a mark of my gratitude, and acknowledgment of the great honor which I have received from the Royal Society.[155]

How far the society went in encouraging original research becomes evident from the fact that Oldenburg devised a means of protecting inventors and securing rights of priority even in unfinished investigations.[156] Pressure was exerted on all the Fellows to do research work, partly perhaps for the selfish reason of having material for the public lectures. The journal-book records: "Ordered, that such of the Fellows as regard the welfare

[154] Weld, *op. cit.*, I, 244 f.

[155] *Ibid.,* p. 245.

[156] Oldenburg proposed "that a proper person might be found out to discover plagiarys, and to assert inventions to their proper authors" (*ibid.,* pp. 329 f.). He made the motion that "when any Fellow have any philosophical notion or invention not yet made out, and desire the same, sealed in a box, to be deposited with one of the secretaries till perfected, this might be allowed, for better securing inventions to their author" (Birch, *op. cit.*, II, 24).

of the society, should be desired to oblige themselves to entertain the society, either per se, or per alios, once a year, at least with a philosophical discourse grounded upon experiments made, or to be made....."[157]

In considering the group of men interested in experimental science, it must not be forgotten that we are dealing with pioneers, and that the opponents of science were in the overwhelming majority. The publications of the society's early members, with their constant defense of the "innocence" of science and of the society, give ample evidence of this fact. Bishop Sprat, the conservative, religious, university-loving historian of the Royal Society, devoted one of the three books of his history to a justification of the society's design; to proving that the new was not always wrong; that experimental knowledge would not interfere with traditional forms of primary education; that it would not shake orthodoxy of belief. Similarly Glanvill, another propagandist for the cause of the society—an enthusiast in science, yet a believer in witches—a Cartesian, although an absolute and devout follower of Bacon's ideas—wrote to defend the society. His pamphlet, *Plus ultra, or Progress and Advancement of Science since the Days of Aristotle. An account of some of the most remarkable late Improvements of Practical, Useful Learning to encourage Philosophical Endeavours, occasioned by a conference with one of the notional way* (1668), was written in reply to Rev. Robert Crosse who dared to maintain that the Royal Society had done nothing to advance science, and that Aristotle's knowledge could not be excelled.[158] In this pamphlet Glanvill reviewed the recent inventions of marvelous scientific instruments, and all this with a

[157] Weld, *op. cit.*, I, 246 f. Cf. also (*ibid.*, p. 249) Oldenburg to Ray: "That the work of the society do not be altogether on the shoulders of three or four Fellows .... you are looked upon as one of those which the Council have in their eye for such an exercise, desiring you that you would think upon such a subject as yourself shall judge proper, etc."

[158] *Ibid.*, p. 229.

"principal eye on the Royal Society and noble purposes of that illustrious assembly, which I look upon as a great ferment of useful and generous knowledge,[159] [which] makes a bank of all useful knowledge, and makes possible the mutual assistance that the practical and theoretical part of physics affords each other. . . . . Indeed, of all combination of men that met for the improvement of science never was any whose designs were better . . . . than the Royal Society.[160] [It] has done more than philosophy of a notional way since Aristotle opened shop.[161]

How much the society needed such defenders was soon made evident by a most virulent attack made upon it by Dr. Stubbe,[162] a physician residing in Warwick, in his book, *A censure upon certain passages contained in the History of the Royal Society* [i.e., Sprat's] *as being destructive of the established Religion and Church of England.* Stubbe even kept up a long correspondence with Robert Boyle trying to win him away from the society.[163] He had a considerable following in Oxford, and "Stubbeite" came for a while to mean the man who would absolutely disapprove of, and work against, the new science. Of such hostile critics there was indeed a considerable number.

The main lines of the activities of the Royal Society have now been indicated; they had all clearly developed within the first decade of its existence. But something must be added in regard to certain early members not yet mentioned, and the society's connection with the publication of Isaac Newton's *Principia.*

In the second decade of the society's existence there came into prominence in the pages of the journal names other than the original Fellows and officers of the society. Newton and Grew became members in 1671; Flamsteed, 1676; Halley, 1678; and these in due time came to be the dominating personalities. Newton is well known, but a few words about the others may be in

[159] Jos. Glanvill, *Plus ultra,* Preface.
[160] *Ibid.*, p. 83. [162] Weld, *op. cit.*, I, 229.
[161] *Ibid.*, p. 90. [163] Birch, *op. cit.*, I, 95.

place here. Nehemiah Grew (1641–1712)[164] had studied in Cambridge and received his medical education and degree in Leyden. His interests were exclusively along the lines of natural science. His first remarkable work was the *Anatomy of Vegetables*, wherein he, simultaneously with Malpighi, insisted on an entirely new, anatomical treatment of botany. His thoughts on this subject grew into the volume *Anatomy of Plants*. Then he turned to the study of zoölogical questions, and in his work, *Comparative Anatomy of Stomachs and Guts*, became the founder of the science of comparative anatomy. In 1672[165] he became a fellow-curator with Robert Hooke; in 1677, successor to Oldenburg as secretary; in 1686 he immortalized himself among the historians of the Royal Society by his *Description of Rarities* mentioned above.

Flamsteed (1646–1719),[166] a self-taught astronomer, later a student of Cambridge, is of special significance in the annals of the Royal Society because as first astronomer[167] of the newly erected Observatory of Greenwich (1675) he represents the first connecting link between this Observatory and the London society, a connection which has persisted to this day. The original cause of this alliance was not so much affiliation of interests as poverty and the necessity of borrowing instruments.[168]

[164] *Dictionary of National Biography*, s.v.

[165] Birch, *op. cit.*, III, 42.   [166] Wolf, *op. cit.*, pp. 454 ff.

[167] Flamsteed did fundamental work in mapping the stars. His memory has been marred by his peculiar conduct toward Newton; he wanted to withhold certain of his observations, urgently needed by Newton; when they were published in spite of him through the Prince of Denmark, he was so furious that he burned three hundred copies of them (Weld, *op. cit.*, I, 378).

[168] The King had left the observatory for a period of nearly fifteen years without a single instrument. Sir Jonas Moore provided Flamsteed with a sextant, two clocks, a telescope, and some books; other instruments had to be furnished by Flamsteed himself, and he in turn borrowed some from the Royal Society. The minutes record: "It was ordered that the astronomical instruments belonging to the society be lent to the Observatory at Greenwich" (Weld, *op. cit.*, I, 255).

134   THE RÔLE OF SCIENTIFIC SOCIETIES

Halley (1656–1742)[169] is among the most remarkable men of the Royal Society and, indeed, of English science.[170] He gained scientific fame by mapping the stars of the Southern Hemisphere from St. Helena, where he was sent at royal expense. In 1678 he became a Fellow of the society, and was sent to Dantzig to settle a dispute between Hooke and Hevelius. He was interested in the study of gravitation, studied earth magnetism, and constructed the isogonal lines of magnetic declination. He calculated the orbit of the "Halley" comet (1682), coupled with a prediction of its return, strikingly verified in 1759.[171] In 1703 he became Savilian professor of geometry in Oxford and in 1720 succeeded Flamsteed as head of Greenwich Observatory.

I shall not attempt to give an account of the work of the society in these later decades of the century, as it proceeded along the lines already laid down, and shall in conclusion only show how close was the connection of the Royal Society with the works of Isaac Newton.[172] Newton, Lucasian professor of mathematics or optics at Cambridge, was proposed as a member by Seth Ward, December 21, 1671, and elected January 11, 1672.[173] He soon communicated the invention of the reflecting telescope. The society was delighted with the report, and Oldenburg at once sent a letter to Paris with a detailed description to secure the honor of the invention to Newton.[174] The following letter

[169] Wolf, *op. cit.,* pp. 463 ff.

[170] He published eighty-one papers in the *Transactions*.

[171] Halley also holds a prominent place in the history of mathematics for his work in the theory of probabilities, in the construction of mortality tables, and solution of higher equations; also for his translation of Greek mathematics from the Arabic (Cantor, *op. cit.,* Vol. III, *passim*).

[172] Newton's relation with the society had commenced in 1669, when he asked Collins to publish a mathematical solution in the *Transactions* anonymously: "For I see not what there is desirable in public esteem where I am able to acquire and maintain it. It would perhaps increase my acquaintance—the thing which I chiefly study to decline" (S. J. Rigaud, *Correspondence of scientific men of the seventeenth century.* . . . .).

[173] Weld, *op. cit.,* I, 232 f.   [174] *Ibid.,* pp. 235 f.

Newton sent to Oldenburg is important for our purpose in showing the real value of the society's assistance to the scientist:

> At reading of your letter I was surprised to see so much care taken about securing an invention to me of which I have hitherto had so little value: And therefore since the Royal Society is pleased to think it worth patronizing, I must acknowledge it deserves much more of them for that than of me, who, had not the communication of it been desired, might have let it still remain in private as it hath already some years.[175]

Newton next communicated to the society his theory of light and colors[176]—in his words: "the oddest, if not the most considerable detection, which had hitherto been made in the operations of nature." It was, according to the letter-book,

> ordered that the author be solemnly thanked . . . . for this very ingenious discourse, and that the society think very fit . . . . to have it forthwith published as well for the greater conveniency of having it well considered by philosophers, as for securing the considerable notions thereof to the author, against the arrogations of others.

These discoveries were the first of Newton's productions which became known abroad. Though he had lectured on his new theories of color for years to students in Cambridge, no report of these had come to the ears even of men so keen to hear as the Fellows of the Royal Society,[177] and it was thus due to their activities that they reached the republic of learned men. This was fully acknowledged and appreciated by Newton, who, in addressing his thanks to Oldenburg, said:

> It was an esteem of the Royal Society for most candid and able judges in philosophical matters that encouraged me to present them with that dis-

---

[175] Rigaud, *op. cit.*, II (January 6, 1671), 311.

[176] A communication from Mr. Newton: "Concerning Newton's discovery about the nature of light, refractions, and colours, importing that light was not a similar, but a heterogeneous thing, consisting of difform rays, which had essentially different refractions, abstracted from bodies they pass through, and that colours are produced from such and such rays, whereof some in their own nature are disposed to produce red, others green, others blue, others purple, etc., and that whiteness is nothing but a mixture of all sorts of colours, or that 'tis produced by all sorts of colours blended together" (Weld, *op. cit.*, I, 237).

[177] David Brewster, *Life of Sir Isaac Newton*.

course of light and colours, which, since they have so favourably accepted of, I do earnestly desire you to return them my most cordial thanks. I before thought it a great favour to be made a member of that honourable body, but I am now more sensible of the advantage; for believe me, Sir, I not only esteem it a duty to concur with them in the promotion of real knowledge, but a great privilege that, instead of exposing discourses to a prejudiced and censorious multitude (by which means many truths have been baffled and lost) I may with freedom apply myself to so judicious and impartial an assembly.[178]

Newton's theory of colors was bitterly rejected and criticized by Hooke and Huygens.[179] As the society communicated this to Newton, it gave rise to acrimonious debate. Newton thereupon wrote that he would never trouble himself again to do experiments;[180] but he continued to publish papers in the *Philosophical Transactions*. In 1672 he asked to resign from the Royal Society.[181] As it was suspected that the weekly shilling might be the cause of this determination, his resignation was not accepted, but he was excused from payments. In 1675 he submitted again to the society his paper on the properties of light, the principal phenomena of colors, and his (wrong) corpuscular or emission theory of light.[182]

In 1684 he came again into the limelight of the activities of the Royal Society never to withdraw until his death in 1725. Indeed the writing and publication of his *Principia*, are intimately connected with the society's work and members. It was noted that Hooke had earlier made extensive studies in the laws of fall-

[178] Weld, *op. cit.*, I, 238.

[179] It should be especially noted that in most of Newton's work in physics his researches were on exactly the same topics as Hooke's (*ibid.*, p. 239).

[180] *Ibid.*, p. 240.

[181] "Since I shall neither profit them nor can by reason of distance [Cambridge] partake of advantage of their assemblies, I desire to withdraw" (Rigaud, *op. cit.*, II, 348).

[182] In the latter he was again attacked by Huygens, and science has of course since decided for the Dutch scientist.

ing bodies.[183] In 1674 the tireless curator published *An Attempt to Prove the Motion of the Earth*, in which he stated that the less the distance of heavenly bodies, the greater their force of attraction. He did not know the law underlying this relation, but suggested that it would be useful to discover it.[184] In private conversation with Halley and Wren, who were also deeply interested in the study of these phenomena, Hooke admitted that he believed the force of attraction to be inversely proportional to the square of the distance—a conclusion Halley had also reached; but Hooke refused to give the mathematical explanation of how bodies subject to that law would move. Upon this, Halley went to Cambridge and asked Newton what the path of a body moving according to this law would be. Newton answered at once that it would be an ellipse and elaborated his explanation and answer in a paper, *De motu*, which he submitted to the Royal Society, April 28, 1686, and which he shortly afterward elaborated into the *Principia*. The Royal Society, to whom this work was dedicated, intended to publish it at its own cost. But when Hooke's claim[185] of priority threatened to delay this pub-

[183] The following statement was made by Hooke in 1665: "Gravity, though it seems to be one of the most universal active principles . . . . has had ill fate . . . . and neglect. The inquisitiveness of the later age has begun to entertain thoughts of it. Gilbert began to imagine it a magnetical attractive power inherent in parts of the terrestrial globe. The Noble Verulam [Voltaire insists upon Bacon as the discoverer of the law of gravitation] also embraced the opinion and Kepler makes it the property inherent in all celestial bodies, sun, stars, planets." But Hooke proposed to try to solve the problem by experimenting (Birch, *op. cit.*, II, 70).

[184] Poggendorff, *op. cit.*, p. 586.

[185] Halley wrote about this claim as follows: "There is one thing more that I ought to inform you [Newton] of, viz., that Mr. Hooke has some pretensions upon the invention of the rule of decrease of gravity being reciprocally as the squares of the distances from the centre. He says you had the notion from him, though he owns the demonstration of the curves generated thereby to be wholly your own. How much of this is so, you know best; as likewise what you have to do in this matter. Only Mr. Hooke seems to expect you should make some mention of him in the Preface, which it is possible you may see reason to prefix" (Weld, *op. cit.*, I, 308 f.).

lication, Halley undertook to print it at his own expense, May, 1687.[186]

These few remarks are sufficient to show that the early development of the scientific career of Newton was most materially influenced by the Royal Society, and that the publication of the *Principia* must be conceived of as a direct result of the help and stimulation both of the society as a whole, and of one of its most important early members, Halley.

The development of the Royal Society, from its informal beginnings to the publication of the *Principia*, has now been reviewed. It has been seen that a body of experimenters and science-loving amateurs, not supported by or affiliated with any learned body—indeed, hardly helped by the King—created a center where the new science could be fostered. We have seen them at work, devising as best they could laboratory facilities, making and improving instruments; experimenting along most varied lines of research; constantly communicating with foreign workers, and establishing the first organ of international scientific communication in the *Philosophical Transactions;* helping the cause of science by encouraging workers and by publishing their works—in short, supplying that most essential aid without which the progress of science would have undoubtedly been delayed for decades. And, what is more important, through their existence and work they made it clear that a new order of things had arisen, that new facts, new methods of work, new interests, were to be recognized in place of the former superannuated and inherited ideas. The Royal Society must therefore be reckoned as first among the pioneer reforming bodies of the century.

[186] An interesting incident in the story of this publication is that Newton wanted to withdraw the third book of his *Principia*, which contains his generalization and application of his mathematical physical laws to the universe and on which his special fame is based: "The third, I now design to suppress. Philosophy is such an impertinently litigious lady, that a man had as good be engaged in law suits, as have to do with her. I found it so formerly, and now I am no sooner near her again, but she gives me warning." It was only out of respect for Halley's wishes and finances that he allowed its publication (*ibid.*, p. 311).

CHAPTER V

THE ACADÉMIE DES SCIENCES

The learned society which came into existence in France in 1666, as has been said above, bore more resemblance to the Royal Society than to the Cimento, but it was in many ways a very different assembly. It was a royal institution not only in name but in deed, and therefore had much of the rigidity of a government institution. But it had also the incalculable advantage of the resources of a royal treasury. While the English body was forever struggling with financial difficulties, the members of the French society drew their fixed pensions, and had the means supplied for experimentation and laboratories. Nay more, the liberal offers of Louis XIV attracted scholarship from all quarters, so that at one time the Academy seemed almost more a continental than a French gathering. Further, the fact that the royal favor came to this learned society from Louis XIV, the greatest monarch of Europe, gave special prestige and glory to the cause of experimental science.

As was said in connection with the Cimento, it is usual to connect the foundation of the Académie des Sciences with the establishment of the famous literary Académie Française. This is in some ways justified, for the precedent established by founding the literary academy in 1635 paved the way for the scientific academy in 1666. This view, however, emphasizes the mere externals of organization, rather than the essential element, the beginnings of co-ordinated scientific work of amateurs in France. Yet it is the study of the latter feature that gives us the key to the origin of the Académie des Sciences, and this will be therefore taken up in some detail.

The earliest instance which has come down to us of scientists meeting regularly to experiment together is the gatherings at the cell of the noted Minorite friar, Morin Mersenne. Mersenne (1588–1648), a man "truly incomparable in his way,"[1] as Boyle calls him, was an investigator of note, mainly in problems of acoustics; but he is of special importance in the history of science because he was a friend and correspondent of most of the prominent scientists of the time, and his correspondence, as is generally admitted, took the place of a scientific journal. He was a most intimate friend of Descartes, and through him Descartes communicated for years (1629–49) with Galileo, Gassendi, Roberval, Hobbes, Carcavi, Cavalieri, Huygens, Hartlib, etc. It was in this correspondence that the method originated—later so generally adopted—of proposing questions and giving prizes for the best solution of problems;[2] here also the student of the history of science finds the first description of many discoveries which were destined to become famous. Mersenne translated Galileo's *Dialogo dei duo massimi sistemi del mondo* into French, in 1634, two years after its condemnation, and later his *Discorsi* on mechanics. As Italian was little understood abroad, Mersenne through these translations may be said to have popularized Galileo on the Continent. It is a fact worthy of notice that this was done by a Catholic priest at the time of the apparent height of the hostility of the church to science.

Mersenne's many popular writings show that he belonged to the Baconian group of men who were bent not only on increasing the mass of scientific truths, but on spreading them among the people. His little book, *Vérité des sciences*[3] (1625), and his

[1] Birch, *Life of Boyle*, p. xxiv.

[2] Rosenberger, *op. cit.*, II, 93. For instance, through this correspondence the famous problem of isochronous pendulum vibrations was proposed and solved by Huygens, which in turn led to his discovery of the pendulum clock.

[3] This is in the form of a dialogue between a Sceptic, an Alchemist, and a Christian Father. The Sceptic is a Cartesian; he doubts all, even mathematics, astronomy where no two men agree, astrology, because nobody knows why Jacob and Esau, who were conceived at the same time, have such different dispo-

tract, *Les questions théologiques, physiques, morales, mathématiques—où chacun trouvera du consentment ou de l'exercise* (1634)[4] are of great interest in this connection. Mersenne's importance in connection with the Académie des Sciences lay in his intelligent appreciation of the works of others and his own skill as an experimenter. His cell became the meeting place of a group of men interested in mathematical and experimental science, eager to communicate their ideas and hear of similar work done elsewhere.

Who were the men who met there? It must first be remarked that France in the middle of the century stands preeminent in mathematics. There was Paul Fermat (1601–65), a parliamentarian of Toulouse, called by Cantor the foremost mathematician of France;[5] the famous mathematician Desargues (1598–1662), founder of the science of descriptive geometry; besides, Roberval (1602–75), professor of mathematics in the Collège de France, whose work was along similar lines as Cavalieri's and Torricelli's. Then there was Blaise Pascal (1623–62), both mathematician[6] and physicist, and the famous

---

sitions! The Alchemist admits that we know only one-hundredth of what we ought to know, and defends the secrecy of alchemistic methods. The errors of Aristotle are enumerated. Bacon is criticized because, in his *Advancement of Learning*, he advocates as new a method which has been adopted. But the author approves of Bacon's idea of experimentation—incidentally regrets that we shall never know how much faster a stone of one hundred pounds falls than a stone of one pound—O, yes, we shall know it in paradise!

[4] This has, in the manner of a modern popular scientific magazine, forty-six questions with answers affixed, containing an incongruous mixture of real science and remarks on religion and conduct—for example: "Why is astrology rejected?" "Why do falling bodies increase in velocity?" "Is it permitted to maintain that the earth moves?" "No, on account of a passage in Joshua."

[5] Cantor, *op. cit.*, II, 798. Fermat was a student of the theory of numbers, and problems of "maximal and minimal values," opponent of Descartes, and one of the pre-Newtonian students of infinitesimals.

[6] He was the rival and foe of Descartes because, continuing Desargues' work, he built up the science of conic sections from a different standpoint. During his frivolous earlier years (1654) a gambler friend had put the question to him, how

Pierre Gassendi (1592–1652), a Minorite, like Mersenne, won to the study of Galileo and Kepler through the famous amateur scientist Peiresc.[7] An opponent of Descartes' teachings, he insisted on upholding empirical against deductive science, and defended the Baconian theory of the importance of experimental research.

These men met at various times,[8] whenever they pleased, at Père Mersenne's cell where they were joined by foreigners interested in science.[9] A good account of these meetings is given by Cassini. A number of savants took pleasure in coming there, and entertaining themselves with astronomical observations, problems of analysis, physical experiments, new discoveries in anatomy and botany. Often they entertained foreigners, among these Oldenburg, and hence, says Cassini, "arose the Royal Society."[10] Later this group of men met at regular intervals (every Thursday) at various houses, including Pascal's.[11] Afterward,

---

gain was to be divided after several winnings. This investigation fascinated him and led to the founding of a new branch of mathematics, viz., the theory of probability. He solved many important problems with which all mathematicians were dealing, for instance, those connected with the cycloid. Through Mersenne's correspondence he heard of Torricelli's investigation concerning the vacuum, and in conjunction with his brother-in-law, Périer, made the famous experiment noted above (chap. ii). With regard to the attitude of the church to the Copernican doctrine, he said: "If the earth moves, it cannot be stopped by papal decree."

[7] See above, chap. ii.

[8] Johann Baptista Duhamel, *Regiae scientiarum academiae historia* (1700), p. 8.

[9] We hear of Sir Wm. Petty belonging to the group; and of Hobbes, who in his eight months' sojourn had daily converse with members of the scientific circle and said that Mersenne's cell was better than a school.

[10] Cassini, "De l'origine et du progres de l'astronomie," *Memoires de l'Académie Royale des Sciences, contenants les Ouvrages adoptées par cette Académie avant son renouvellement en 1699*, V, 26.

[11] Jöcher *Allegemeines Gelehrten Lexicon*, m. 522. Gatherings of men more interested in physics and medicine were held at that time in the house of the French physician Abbé Bourdelot (1610–85). Gallois (1673) published *Conversations de l'Académie de M. Abbé Bourdelot*.

THE ACADÉMIE DES SCIENCES

some of these mathematicians and physicists met regularly for four or five years, at the house of Hubert de Montmort, member of the Council of State and *maître des requêtes*.[12] He seems to have been a man of the Peiresc type, friend of Gassendi, but adherent of the Cartesian doctrine; of his scientific accomplishments it is only known that he was skilled in dissections.[13] A report of these meetings by Tuke to the Royal Society gives the impression of a definite organization, and program, and shows that their main interests were in experimentation.[14] Tuke was taken by Roberval to this assembly, "whose business is to advance knowledge of nature by conference and experiment." "There I found near twenty persons sitting in a semicircle about the table. .... As soon as we were seated, M. de Sorbière, secretary to the assembly, addressing the president, told that the gentleman who was to have spoken was ill. .... So they fell to discourse of an experiment which they had lately made." He relates that upon his telling of Boyle's experiment "they desired me to send them Boyle's book, with the promise if they made any discoveries worthy of our knowledge, they would freely impart them to us."[15] The Danish anatomist, Stenon, performed dissection before them, and the experiment of transfusion of blood was tried.[16] The most momentous scientific event in the history of this academy must have been when Chapelain read a letter of Huygens,[17] telling about the rings of Saturn. Huygens in consequence was introduced to the society and in 1663 specially asked to be present.

This assembly was in correspondence with the Royal Society in 1660, and evidently tried to duplicate its work on French

[12] Charles Adam, *Philosophie de François Bacon* (quoting Lettres de Jean Chapelain), p. 338.
[13] *Ibid.*, p. 339.
[14] Birch, *History of the Royal Society*, I, 26.
[15] *Ibid.*, p. 27.   [16] Adams, *op. cit.*, p. 339.
[17] J. Bosscha, *Christian Huygens; Rede am zoosten Gedächtnistage seines Lebensendes* (übersetzt), p. 27.

soil.[18] This is clear from a letter of Huygens who wrote that Montmort's academy, "anxious to emulate the London society," would again apply itself to experiments.[19] But it was felt that there was a great difference between the situation in England and in France. In France, owing to unhappy conditions, the nobility and gentry had no leisure to cultivate their minds by letters, while the Royal Society, as it seemed to Montmort, was mainly recruited from these groups.[20] Another difference was that while the Royal Society prided itself on "owning no hypothesis," the French assembly tried to propagate Cartesian views and principles.

Later, the scientists met at the home of Melchisedec Thevenot (1620–92),[21] a man of rare "curiosity," who had studied everything—history, geography, mathematics, physics, philosophy, and languages—and had an extended correspondence. Here, as at Montmort's, Stenon performed his famous dissection,[22] and Cartesian views were propounded. What the scientists accomplished in these gatherings can be surmised from the early papers published by the Académie des Sciences.

Colbert knew of these meetings at Thevenot's through Perrault and others, and proposed to Louis XIV to give these informal gatherings of scientists an official status, such as the Académie Française and the Royal Society already enjoyed. According to Duhamel,[23] he first planned meetings of three groups: mathematicians and physicists, Wednesday and Saturday; historians, Monday and Thursday; literary men, Tuesday and Friday; and on one Sunday of each month there was to be a joint meeting. But this plan came to nothing, as the historians were too much interested in church history, and the literary class

[18] Birch, *History of the Royal Society*, I, 49.
[19] *Ibid.*, p. 27.   [20] Adam, *op. cit.*, p. 338.
[21] He was famous as a great traveler and author of the work *Relation des divers voyages curieux* (1663).
[22] Foster, *op. cit.*, p. 106.   [23] Duhamel, *op. cit.*, p. 3.

seemed to suggest rivalry with the Académie Française, so the plan reduced itself to arranging for meetings of mathematicians and physicists only. Colbert's aim was to help, thereby, both theoretical science and those investigations which would ultimately advance the *arts et métiers* of France.[24] That he was its spiritual father rather than Louis XIV is evident from the fact that the Academy flourished only during Colbert's lifetime, until 1683, then declined, until in 1699 it was brought to a new life by an entire reorganization. My account will therefore deal separately with the features of the Academy under Colbert's régime and those during the interregnum following his death; it will only touch upon its reorganization in 1699, as its activities thenceforth fall outside the compass of this investigation.

The first appointees[25] of Colbert were men of very different

[24] *Ibid.*

[25] They are as follows: J. Bertrand, *op. cit.*, pp. 3 ff., and Maindron, *Révue scientifique de la France et de l'etranger* (1881), p. 685.

Auzout, prominent astronomer, and inventor of the telescopic micrometer.

Bourdelin, chemist.

Buot, engineer and instructor of pages. Least significant member. He had been a workman and knew no word of Latin.

Carcavi, geometer; he had held position as *conseiller* of the Parlement de Toulouse, in which post he was successor of Pierre Fermat and his scientific executor; in 1666 held the position of librarian of the King's library, was distinctly a non-professional person, and responsible for meetings.

Couplet, junior member; professor of mathematics at the Collège de France and student of mechanics.

Cureau de la Chambre, physician to Louis XIV, also member of the Académie Française.

Delavoye Mignot, junior member, geometer.

Dominique DuClos, chemist, physician of Colbert; he was an alchemist, but seeing his folly before his death burnt his writings; one of the most active members.

Duhamel, anatomist; secretary, on account of his good Latinity.

Frenicle de Bessy, geometrician; a magistrat and author of work on magic squares and theory of numbers.

Gayant, anatomist; helper of Perrault.

Abbé Gallois (1632–1707), later professor of Greek and mathematics at the Collège de France, friend of Colbert; successor of Denis de Sallo in editing the

station in life, physicians, engineers, parliamentarians, all with pronounced interest in science. Three astronomers, three anatomists, one botanist, two chemists, seven geometers, one mechanic, three physicians, and one unclassified member were in this group. It is not clear what motives controlled these appointments. Duhamel[26] suggests that Colbert picked out not scholars, but skilled men, whom every kind of study delighted,

---

Journal des Sçavans (1665-74); intimate with Colbert, telling him of doings of Academy (Fontenelle, *Oeuvres de Fontenelle*, ...).

Christian Huygens (1629-95), the only foreign member among first appointees. He was Dutch, but had made his home for some time in France; mathematician, astronomer, physicist, designer and maker of instruments (air pump and telescope), not connected with any university, but typically amateur and non-professional, he is one of the most characteristic figures of seventeenth-century science. When Colbert appointed him in 1666, he was very prominent in science, had improved the telescope, had discovered two new moons of Saturn, had enunciated the theory of the rings of Saturn, and had done great work in mathematics. He had been elected Fellow of the London Royal Society. It is evident that Colbert applied to him for advice (Bertrand, *op. cit.*, p. 9) about the work to be done by the Academy, from a manuscript letter of Huygens which is extant and has the comment *bon* of Colbert on the margin. That he was most closely affiliated with the Academy is evident from the fact that for twelve years of his sojourn in Paris he lived in the building where its meetings were held (Bosscha, *op. cit.*, p. 29; Foster, *op. cit.*, p. 50).

Marchand, botanist; head of the royal garden.

Mariotte, physicist; overshadowed by the genius of Huygens; he was one of the famous scientists among this body.

Niquet, junior member; geometer.

Pecquet, anatomist; discoverer of the thoracic duct and of the circulation of blood in foetus; a man of highest prominence in the history of physiology (Foster, *loc. cit.*).

Perrault, the most active member; notable as an architect, builder of one colonnade of the Louvre. He took up successively many sciences—anatomy, zoölogy, physics, mechanics; conservative in his views, but an indefatigable worker. It was he who interested Colbert in science.

Picard, astronomer; also one of the best workers; friend of Gassendi; professor of astronomy in Collège de France.

Pivert, junior member; astronomer.

Richer, junior member; astronomer.

Roberval, the mathematician of the Mersenne Academy.

[26] Duhamel, *op. cit.*, p. 4.

and who specialized in one;[27] moreover, men who were attached to no sect.[28]

If the assembly thus established had any statutes or rules, nothing is known of them;[29] no document giving any enlightenment on the subject exists in the archives. A room in the King's library where learned books were kept was assigned to them; the apartment adjoining was to be the laboratory.[30] December 22, 1668, Carcavi reported to the assembly the design with which the King had convoked them. Then the question came up—Should the physicists and mathematicians hold their meetings together on Wednesday and Saturday? It was argued that those who excelled in mathematics were versed in natural philosophy, and that geometric exactness would be of great value to the non-mathematical scientists, to keep them from fallacious opinions; that the separation of physics from mathematics had been the cause of its sterility. The fact that Galileo, Descartes, and Gassendi had evidently belonged to both groups proved conclusive. So it was decided that they would always meet jointly, but that on Wednesday, geometrical, and on Saturdays, physical problems should be discussed. All sessions were to be secret, so as to avoid the possibility of plagiarism, and secure due credit to the inventor.[31] Strangers could only come for the purpose of showing some new thing—hence the scanty reports

[27] Later there were added to the first appointees: Blondel and Cassini (1669); Roemer (1672); Dodart (1673); Borel (1674); G. J. DuVerney (1674); Leibniz (1675) (?); P. de la Hire (1678); Sedileau (1681); Tschirnhausen (1682); Polheuse excl. (1682); Lefevere excl. (1682); De Bessé (1683); Mery (1684); Thevenot, Rolle, Cusset (1685); Varignon (1688); Tournefort (1691); Homberg (1691); Charas (1692); De la Coudray, Morin (1693); Cassini, P. de la Hire, Boulduc, Maraldi, De Chazelles (1694); Fautel de Lajai (1696); Sauveur, Guglielmini (1696); Fontenelle, Carbé, Tauvry (1697); Langlade (1698); Lémery (1699).

[28] There were at first neither Cartesians nor Jesuits among them. This may be the reason why Thevenot, a Cartesian, was asked to join only in 1685.

[29] *La Grande Encyclopédie.*

[30] Duhamel, *op. cit.*, p. 5.   [31] *Ibid.*, pp. 6 ff.

of the sessions. The *pensionnaires* received one thousand five hundred livres from Colbert, and were to give all their time to the society. Besides the King established a fund of twelve thousand livres for expenses, instruments, and new inventions, and to supply the laboratory; thus giving the necessary financial support.

The method of work of the French scientists was different from that of the Royal Society.[32] The experiments were chosen and discussed in advance and then performed in the laboratory next to the library. The members actually worked together; they did not do their work in their homes, and bring reports or repeat experiments in the sessions. In the laboratories the experimenting and observations were jointly made,[33] hence most of the work was reported as the joint product of three or four workers; discussions in regard to results obtained through collective experiments and simultaneous researches were a feature of the meetings. About these sessions we find it stated that one experiment at least was exhibited;[34] that they were devoted to one question, which often remained on the docket for several weeks; that each difficulty was weighed and discussed, and experiments to settle opposing views were made. Such sessions were always full of interest, if not of great benefit to science.[35]

The question of what line of work was to be pursued presented itself immediately. Duhamel has it that from the beginning the fact was emphasized that the inventions of others should be tested, instead of confining the association's attention to making new discoveries.[36] Each academician was asked to

[32] *La Grande Encyclopédie*, s.v., "Académie des Sciences."

[33] Louis F. A. Maury, *Les académies d'autrefois. L'ancienne Academie des Sciences*, p. 15.

[34] Bertrand, *op. cit.*, p. 18.

[35] *Ibid.*, p. 31. How persistent the workers were at times is shown by the fact that in 1669 for twenty successive Saturdays (owing, says Duhamel, to the talkativeness of DuClos) they discussed the same question of coagulation.

[36] Duhamel, *op. cit.*, p. 7.

submit a program of work.[37] It can be surmised that the reports of the sixteen members were essentially different—*quot capita tot sensus*.

The idea of Huygens, or rather of Bacon, of jointly compiling and amassing facts for one great work, which we do not meet in the annals of the Royal Society, was carried out in some instances by the Academy, and gave rise to one of its most successful achievements—its history of animals and plants. Perrault, struck with the existing ignorance of the nature of plants and animals, advised their systematic study and description, with emphasis upon their different anatomical features.[38] He dissected a great many animals, especially the strange ones he could get from the menagerie of Versailles.[39] Such dissections were made with the greatest care. It took three sessions to dis-

[37] Bertrand, *op. cit.*, p. 8. The following was that of Huygens, surely the one most carefully considered:
1. Experiments with vacuum and determination of the weight of air.
2. Examination of force of gunpowder.
3. Examination of force of water vapor.
4. Examination of force and swiftness of wind, and use for navigation and machines.
5. Examination of laws of percussion.

"The main occupation and most useful of this assembly ought to be to work up a Natural History after the design of Bacon. . . . . To know what weight, heat, cold, magnetism, light, color is; of what parts air, water, fire and other bodies be composed; what is the use of respiration of animals; how metals, stones, herbs grow; of all this nothing or very little is known, yet there is nothing the knowledge of which would be more desirable or useful. Under each of these headings the experiments are to be collected, *not* those rare or difficult, but those most essential for the research. . . . . Chemistry and dissection of animals are necessary, but ought to be employed only in so far as they augment the Natural History. . . . ."

[38] Duhamel, *op. cit.*, p. 10. In his program of work, he pointed out that the uses of many parts of the body are not known and should be investigated by dissection; that in botany they should observe with the microscope the changes undergone by seeds; what the saps contain, what salts are in the ash; whether nutrition and growth in plants are similar to those of animals.

[39] Bertrand, *op. cit.*, p. 14. The story goes that he was surprised by the King at the dissection of an elephant, and could not at first be found by his royal visitor, on account of the huge size of the dissected object.

sect the trunk of the elephant. "He thus," says Daremberg, "advanced the physiology of the senses and the study of animal mechanics, described the peristaltic movement of the entrails, and rectified the idea of the bile vessels of animals." Perrault was specially anxious to make experiments with a view of dispelling absurd popular prejudices. He tested the chameleon—and found no change of color; he experimented with the salamander—and did not find it incombustible; he discovered that the pelican did not nourish its young with blood—"thus systematically he eradicated misconceptions," says Condorcet.[40] These researches were published as *Mémoires pour servir à l'histoire naturel des animaux* by the Royal Press, and translated into English by order of the Royal Society.[41]

The Preface to the book emphasizes the fact that this type of work could have been done only by co-operation:

> That which is most considerable in our Memoirs is that unblemishable evidence of certain and acknowledged verity. For they are not the work of one private person, who may suffer himself to be prevailed upon by his own opinion; who can hardly perceive what contradicts his first conceptions . . . . these memoirs contain only matter of fact, verified by whole society.

Another work which represented the joint effort of many members was the *Mémoires pour servir à l'histoire des plantes*, published by DuClos, Perrault, Gallois, Bourdelin, Dodart, and Borelli. How meager the botanical knowledge was with which the Academy started can best be proved by some of the botanical questions of Perrault:

> Is it true that a plant can reproduce itself of salts taken from its ash?
> Does the earth reproduce plants by its own fecundity, without seed?

---

[40] M. J. Condorcet, *Oeuvres complètes de Condorcet*, Vol. I. *Eloges*. Perrault.

[41] "The Natural History of Animals, Containing the anatomical description of several Creatures, dissected by the Royal Academy of Science at Paris, wherein the Construction, Fabrick and genuine use of the Parts are exactly and finely delineated in Copper plates and the whole enriched with many curious Physical and no less useful Anatomical Remarks, being one of the most considerable productions of that Academy" (London, 1702).

Does there exist in plants as in animals a soul which causes its movements—is it the root?

Are there sympathies and antipathies among plants?

Great efforts were made by Dodard to examine plants chemically. He tirelessly distilled saps, extracted oils, but as he threw away the ash as *caput mortuum,* he arrived at the same analysis for most diverse vegetation![42] A close study was made by Mariotte, Perrault, and DuClos to discover whether something similar to the circulation of the blood could be found in the flow of the sap of plants. The volume which resulted from these researches was so well received that three years later a second edition was needed.

A third instance of co-operative effort was a *Treatise on Mechanics.*[43] This was, however, of no scientific value.

Yet it would be an utter mistake to assume that the academicians in these early years carefully planned a series of investigations for building up systematically the various branches of science; a few instances of such contributions have been selected to show the possibility of the development of such a method.

To turn now to the other most distinctive feature of the Academy's first years we find that astronomical research was, from the first, of great interest to the academicians, and that soon after the establishment of the Academy the astronomers met in the garden of the King's library in order to make observations in common. Within one year, at the request of Auzout and Picard,[44] the cornerstone of the Observatory of Paris was

---

[42] Maury, *op. cit.,* p. 16.   [43] Bertrand, *op. cit.,* p. 13.

[44] In 1664 Auzout and Picard had written to Louis XIV saying: "It is a misfortune that there does not exist in Paris, nor anywhere in your kingdom, an instrument with which I can determine precisely the height of the celestial pole" (Wolf, *op. cit.,* p. 449). Birch, *History of the Royal Society,* II, 237. Oldenburg claimed that Auzout, on reading Sprat's *History of the Royal Society,* asked for new astronomical instruments, which would be an interesting instance of the interaction among the societies.

laid,[45] and put into the complete charge of the Academy.[46] The phase of work connected with this Observatory is one of the most notable in the history of the Académie des Sciences. Upon the advice of Auzout and Picard, Cassini was called from Italy in 1669 to fill the post of head astronomer, and he and his family were destined to make the Paris Observatory the foremost in the world.[47]

There was another type of scientific enterprise which the Académie des Sciences could afford to undertake, but which was never within the means of the Royal Society. Within the first years of its existence, it sent out two scientific expeditions. One was to Uranienburg, where Picard wanted to test Tycho de Brahe's observations, and where he actually did find a very essential error in Tycho's calculations of the meridian. From this expedition he brought back the famous Danish astronomer, Roemer, to Paris.[48] The other, fraught with important consequences for science, was the expedition to Cayenne, to make astronomical observations near the Equator. From there Richer reported the fact, considered most puzzling, that the second pendulum did not vibrate at the same rate at Cayenne as in Paris. It was surmised that the difference was due to the temperature, but in time the spheroidal shape of the earth was recognized.[49]

[45] Perrault designed the Observatory, with more consideration, it is said, of architectural beauty than for its adaptation to astronomical observations (Bertrand, *op. cit.*, p. 22).

[46] It is to be observed that while Louis XIV put the Observatory directly into the charge of his Academy, the Greenwich Observatory (1675) was a foundation apart from the Royal Society.

[47] Cassini himself is an interesting type of the seventeenth-century personality; an astrologer, he is "cured" by studying comets, seeing that even their movements are subject to laws, and hence cannot be prophetic of catastrophes and calamities. The most spectacular of his astronomical achievements was the discovery of four of the eight moons of Saturn (Maury, *op. cit.*, p. 22).

[48] *Ibid.*, p. 21.

[49] *Ibid.*, p. 31. An account of these travels was published in 1693 in a beautiful quarto volume: *Recueil d'observations faites en plusieures voyages par*

THE ACADÉMIE DES SCIENCES 153

In England an effort was made, very soon after the installation of the society, to issue periodic publications. Such was not the case in France, mainly because the *Journal des Sçavans*, which for a while was edited (1665-74) by one of its members, Abbé Gallois, supplied the outside world with the main occurrences in the society.[50] But the work of the early years has since been collected—to a great extent from the pages of the *Journal des Sçavans*—and published in eleven volumes of the *Histoire de l'Académie des Sciences depuis son établissement en 1666 jusqu'à 1699*. Glancing through them, we get an impression of most varied scientific work.[51] The earliest efforts under the lead of Huygens, Picard, and Auzout dealt with the perfecting of the telescope. Then we have experiments turning on such questions as to whether butter melts and plants grow in a vacuum. The discovery was made that in a vacuum some air is exhaled —apparently what was contained within the substances; a fish in water was put under the receiver and died because air was exhausted from its swim-bladder. The use of the telescope for terrestrial purposes, to measure angles, was noted here for the first time.

Very interesting, from a historical point of view were the experiments Huygens and his assistant, Papin, made (1673) with the cylinder, using gunpowder as the motive force, "pour avoir toujours a son commandement un agent très puissant et qui ne coûte rien à entretenir comme font les chevaux et les hommes."[52] Human dissections were demonstrated by Pecquet and Perrault, and studied with the greatest possible thoroughness. The question of the transfusion of blood deeply stirred the

---

order de sa majesté pour perfectionner l'astronomie et la geographie—par messieurs de l'Académie Royale des Sciences.

[50] See chap. vii.

[51] No attempt will be made at exact references, as the subjoined matter is merely selected at random.

[52] Bosscha, *op. cit.*, p. 35.

company, and the experiment was tried on dogs, until a decree of the courts forbidding this method put an end to these investigations.[53] The chemical experiments of DuClos and Bourdelin, for instance, in regard to the mineral salts contained in the waters at places like Vichy, fill many folio pages of the volumes. The mathematicians, Roberval, Mariotte, Perrault, Frenicle, studied the problems of free fall and the periodicity of the comet; Perrault studied the ear; Mariotte the phenomena of air-pressure, the theory of the colors in soap bubbles and in the rainbow. The properties of phosphorus were investigated, and in 1682 the (Halley) comet was, of course, observed with intense interest. In these annals of the Academy during its first years are some of the most famous facts in the world of science, as, for example, Mariotte's discovery of the blind spot in the eye; Roemer's calculation of the velocity of light from the satellites of Jupiter; Huygens' undulatory theory of light explaining the double refraction observed in Iceland spar by Erasmus Bartholin.

There is repeated mention of people not connected with the society sending or presenting in person books to be read and experiments to be tested.

Turning from the volumes of the *Histoire* to activities of the Academy not recorded there, we note that the Academy has the distinction of having published under its auspices one of the epoch-making scientific books of the century, namely, Huygens' *Horologium*,[54] which contained the description of the pendulum clock, second in importance in astronomy only to the telescope. In the Dedication addressed to Louis XIV, the author expressly

[53] See above, chap. iv. A further evidence of the interest of the Academy in medical matters is that Colbert gave the academicians the right to visit incurables at l'Hôtel Dieu—but the nuns there did not permit it (Bertrand, *op. cit.*, p. 15).

[54] C. Huygens, *Opera varia*, p. 19. The topics are the following: "1. Pendulum clock. 2. Fall of bodies. 3. Motion along cycloids. 4. Evolutes, Centre of oscillation. 5. Centrifugal force. Cycloid (Evolute of cycloid)."

thanked the King for the opportunity which had been provided for his researches by supplying him with a well-equipped observatory and with leisure for work.[55]

It remains to emphasize the practical side of the efforts of the first years of the Academy, and those activities which kept it in contact with the world outside of its laboratories. In 1666 Auzout proposed as his program of work that a commission be appointed to investigate the methods of artisans and to study their utensils and instruments and such defects as they might have.[56] Roberval, Mariotte, Roemer, and Blondel were much interested in mechanics, and Blondel read, each week, to the company a description of a machine.[57] A collection of tools, machines, and instruments was started, and it became customary for people making improvements and new inventions to submit them to the Academy for its approval.[58] Such tools and instruments as it examined and approved of were incorporated into its collection. Indeed the King ordered them to occupy themselves mainly with a description of mechanisms. Binot and Couplet started a catalogue[59] which contained descriptions and pictures of an infinite number of devices suggested by members of the Academy and by outsiders—cranes, cylinders, and machines to raise water. Perrault and Huygens are conspicuously represented among the contributors, and many optical instruments and models of Cassini are found in its pages.

[55] Other books were published by the Academy in its first decade, and we hear of their being shown to the King in 1681 (Duhamel, *op. cit.*, Preface).

[56] Bertrand, *op. cit.*, p. 8.

[57] Duhamel, *op. cit.*, p. 162.

[58] Cusset, for instance, showed them a model of an engine to raise water (1698), and we hear of Leibnitz submitting his calculating machine to the society (1675).

[59] It was published as *Machines et inventions approuvées par l'Académie royale des science, depuis son établissement jusqu'à présent; avec leur déscription. Dessinées et publiées du consentement de l'Académie* (7 vols.), Vol. I (1666–1701).

Another task combining scientific investigation and practical usefulness was put upon the Academy. The King ordered a map of France to be made, and this task fell to Picard and Philippe de la Hire, a new and most versatile member of the Academy. Its execution necessitated De la Hire's visiting Bretagne, Guienne, and La Provence, from which he brought back many objects and observations of interest for the society.[60]

During these years the personnel of the Academy had somewhat changed. The accession of the two brilliant foreigners, Roemer, from Denmark, and Cassini, from Italy, has been mentioned; also the membership of Philippe de la Hire, who became one of the most prominent and characteristic men of the assembly.[61] There must however be registered the loss of Huygens and Roemer due presumably to the revocation of the Edict of Nantes. In 1682 several foreign scientists joined, becoming only associate members because of their unwillingness to expatriate themselves: Viviani, Steno, Hartsoecker, and Tschirnhausen.[62]

As was said above, with the death of Colbert (1683) there set in a period of decline for the Academy. The man who was instrumental in the revocation of the Edict of Nantes, Louvois,[63] took Colbert's place, a man who had no sympathy with scientific work. For a while the academicians were degraded into serving merely the personal curiosities of the King and state; Roberval's mathematics in assisting games of chance, Mariotte's hydrostatics for the Versailles cascades (1684–1713), Blondel's mechanics mainly for the purposes of artillery. The whole ten-

---

[60] Maury, *op. cit.*, p. 25.

[61] De la Hire was a mathematician and his work on conics, continuing Desargues' researches, was almost as revolutionary as Descartes', and is in line with the most modern treatment of the subject. Like Perrault, he was interested in the most diverse branches of learning, and very conspicuous as one of the founders of meterological observations.

[62] Maury, *op. cit.*, p. 23.   [63] *Ibid.*, p. 37.

THE ACADÉMIE DES SCIENCES

dency was to make the work of the Academy more practical. In 1686 Louvois sent word that he did not wish so much interest in "recherche curieuse, ce qui n'est qu'une pure curiosité ou qu'est pour ainsi dire un amusement des chimistes," but attention to "recherche utile, celle qui peut avoir rapport au service du roi et de l'Etat."[64]

That the meetings of the Académie were from the beginning far from peaceful is brought out by all historians.[65] In fact, they became the battle ground between Cartesianism, hallowed by the conservative older element, especially supported on French soil on account of Chauvinistic reasons, and anti-Cartesians, soon to be Newtonians.[66] Cartesianism was the issue between the two schools of physiologists, between the two classes of physicists, and in a much graver form between the two classes of mathematicians. Rolle, De la Hire, Tschirnhausen, and Gallois resisted the introduction of Newtonian calculus in opposition to l'Hôpital and Varignon—a struggle ending after a battle of five years in the victory of infinitesimal calculus.[67] Among the members petty jealousies had arisen, and whereas the original purpose had been, as in the Cimento, to publish the work of the Academy as a unit, they had drifted so far from this ideal that in 1688 they asked that a commission be appointed to investigate the works to be published by members, and determine whether the author was guilty of plagiarism.[68]

From 1688 to 1691 there seem to have been not enough experiments to keep a two hours' session busy, and accounts of the discussions show them to have been meager. Yet some matters may be of interest and worthy of note as indicating the cosmopolitan range of the Academy, e.g., the Jesuit Fathers sent in

[64] Bertrand, *op. cit.*, p. 40.

[65] Maury, pp. 26 ff.

[66] Charles Adam, *op. cit.* (p. 342), says the members of the Academy believed they were Cartesians, but were Baconians, and used the same methods as the Royal Society.

[67] Maury, *op. cit.*, p. 64.   [68] Bertrand, *op. cit.*, p. 45.

observations of the heavens from China; the Siamese ambassador brought a Siamese astronomy which Cassini deciphered. In 1690 the ex-king, James II, visited the Observatory of Paris and discussed Newton's idea of the spheroidal shape of the earth.[69] The fact that he was shown the place where machines and burning mirrors were kept indicates the pride of the society in these possessions.

In these years, two men were appointed to the Academy who were especially to emphasize experimentation along lines of chemistry.[70] In 1699 Nicholas Lémery (1645–1715) became a member—to my mind, one of the most interesting men of the period. The pupil of an apothecary, really self-taught, he condemned the alchemistic obscurities which clung about chemistry, and delighted in teaching it in a simple, straightforward way, with experiments, in his own room, "qui était moins une chambre qu'une cave, et presque un antre magique, éclairé de la seule lueur des fourneaux."[71] So many people—even ladies—attended these gatherings that there was hardly room for the experiments. He gave up the Calvinistic religion to join the work of the Academy. Here he conducted courses in chemistry for large audiences of working people, the only instance of teaching connected with the Academy I find recorded. To Lémery is due in great part the love for chemical experiment as such, the popularization of the science, and bringing it within the reach of the average student and man.

The other man was Homberg (1652–1715)[72] a Dutchman born in Java, by profession a lawyer, who, "tired of the arbitrary laws of man, looked for the laws of nature," and was won

[69] *Ibid.*, p. 37.
[70] Maury, *op. cit.*, p. 40.
[71] Fontenelle, "Lémery," *op. cit., Eloges*. He lived to see his book, *Cours de chymie*, in its thirteenth edition and translated into all the languages of Europe. "It is a practical book, not dealing with theories, and sold," says Fontenelle, "just like a novel or satire."
[72] Fontenelle, "Homberg," *op. cit.*

THE ACADÉMIE DES SCIENCES

to experimental science by the experiments of Guericke. He visited many universities, worked in the laboratories of Boyle, and studied medicine and anatomy with de Graaf in Leyden. His reputation as a chemist brought him into relation with the Duke of Orleans,[73] who drew him to his chemical researches under the official title of "physician." In 1691 Homberg was appointed director of the chemical laboratories of the Académie des Sciences, and was responsible for what was shown at the sessions, a position somewhat like that of the curator of the Royal Society.[74]

The years 1692–99[75] form a separate chapter in the history of the Academy; Pontchartrain, taking Louvois' place, put the Academy under his nephew, Bignon, a Maecenas of science, whose interest in the Academy culminated in its complete reorganization in 1699. In that year the society received the constitution under which it was to exist until its dissolution in 1793. Its statutes[76] in a way represent what the experience of thirty-three years had taught the body as the best line of co-operative

[73] See above, chap. ii.

[74] Homberg's experiments bore on questions of heat, magnetic declination, and the chemistry of plants; he was also deeply interested in questions of alchemy (Haeser, *op. cit.*, II, 437).

[75] In this period the following prominent scientists joined the Academy:

L'Hôpital, a mathematical genius, who died at the age of forty. He was among the first to understand the vast importance of infinitesimal calculus, first interested and taught calculus to Huygens, and wrote the first textbook on this subject. His extensive correspondence with Huygens and Leibnitz is of great importance.

Méry (1645–1722), prominent in the history of physiology for work on the ear, and circulation of blood in the foetus; also for the emphasis he laid on surgical teaching.

Littré, an indefatigable anatomist, who had in 1684 made more than two hundred dissections.

Tournefort (1656–1708), a prominent botanist. He became head of the Jardin des Plantes in 1683, traveled extensively in the Orient, and is recognized as an important forerunner of Linnaeus (Maury, *op. cit.*, p. 5).

[76] Published in *Œuvres de Fontenelle*, I, 63–72.

work, and thus, for us, are of more interest as a résumé of past than as a beginning of future work.

The body was increased to fifty members: ten honorary members; twenty *pensionnaires,* the real workers, inhabitants of Paris, and distributed through the various branches of science as follows: three for geometry, three for astronomy, three for mechanics, three for anatomy, three for chemistry, three for botany, one secretary, and one treasurer.[77] Then there were to be twenty associates, twelve resident in Paris—two attached to each branch of science—and eight[78] foreigners. Besides, provision was made for twenty pupils, each attached to one *pensionnaire.*

I shall quote some of the most characteristic statutes of the newly organized Academy:

XII.[79] No man who is a priest . . . . or belongs to any religious order, shall be proposed, except as an honorary member.

XIII. Only those persons shall be proposed to the King for the place of pensionnair or associate that have distinguished themselves by some considerable published work, by some course whereby they made a reputation in the science they professed, or by some machine of their own invention, or by some particular discovery.

XX. It having been found by experience that there are disadvantages in the tasks to which the academicians apply themselves in common, each one shall choose a particular object for his studies, and by the account he shall give of it in the meeting, he shall endeavor to enrich the Academy by his discoveries and improve himself at the same time.

[77] The *pensionnaires* nominated were in great part the academicians of previous years: mathematics, Gallois, Rolle, Varignon; astronomy, Cassini, De la Hire, Le Fevre; mechanics, Filleau des Billets, Jangeon; anatomy, Duhamel, Du Verney, Méry; chemistry, Bourdelin, Homberg, Boulduc; botany, Dodart, Marchant, Tournefort; secretary, Fontenelle; treasurer, Couplet. Lémery was an associate in chemistry.

[78] The eight foreign associate members were Leibnitz, Tschirnhausen, Guglielmini, Hartsoecker, Bernouilli (Jacob and Johann), Roemer, Newton, Viviani (later) (Joseph Chamberlayne, *The Lives of the French, Italian and German Philosophers, late members of the Royal Academy of Sciences in Paris,* p. 99).

[79] The numbers refer to the statutes.

THE ACADÉMIE DES SCIENCES

XXII. Though each academician be obliged to apply himself to a particular science . . . . everyone shall be exhorted to extend his inquiries into all that may be useful and curious in the several parts of Mathematics, in the different divisions of the Arts, or in that which may in any way relate to the knowledge of Natural History and Philosophy.

XXIII. At every meeting, two at least of the pensionnairs shall be obliged to bring some of their particular observations. . . . .

XXIV. All observations . . . . shall be . . . . delivered in writing to the Secretary.

XXV. All experiments reported shall be repeated if possible at the meeting, otherwise at home in private, with some of the academicians present.

XXVII. Correspondence with learned men is to be cultivated.

XXVIII. The Academy shall appoint some of their members to read all valuable articles that appear on Physics and Mathematics.

XXIX. The Academy shall repeat all considerable experiments and note . . . . conformity or difference between their own and other observations.

XXX. If a member proposes to print a book, it shall be . . . . submitted to the examination of the Academy. . . . . None of the members shall put the title of *Academician* at the head of any work, unless it has first been approved by the Academy.

XXXI. The Academy shall examine . . . . all machines for which patent and privilege is desired by inventors . . . . and inventors of such as have been approved shall be obliged to leave a model in the Academy.

XXXV. The public shall be admitted to two open meetings.

XLVII. His Majesty will continue to pay ordinary pensions.

XLVIII. . . . . the King will continue to allow necessary charges for making the several experiments and discoveries of every academician.

The proceedings of the Academy, comprising the entire memoirs of the scientists, were to be published yearly; thenceforth this society had its periodic publication like the Royal Society.

The Academy received new apartments in the Louvre, where, after the reorganization in February, 1699, it was formally opened in public session on April 29, 1699. Bignon, in opening, well characterized the task of the Academy of Sciences:

L'Académie Française avait pour son partage l'art de la parole avec tous ses agréments, mais l'Académie des Sciences n'aspirait qu'à la vérité et souvent la vérité la plus sèche et la plus abstraite, qu'il lui suffisait que le vrai put être utile et qu'elle le dispensait d'être agréable.[80]

Fontenelle prefaced the first volume of his history of the Academy since its reorganization with an essay somewhat prophetic in its contents, which clearly shows that the thirty-three years of labor of the academicians were felt to be of great value to science and to men in general. He feels convinced that this new science, studied by such a small number and hardly noticed abroad, is destined to give great advantages to innumerable people; better anatomy has given rise to more skilful surgery, and will do so to a greater extent in the future; cycloids, though first studied for "vanity," have given the pendulum clock to the world (Huygens); comparative anatomy will ultimately help man, since in the mechanism of one species of creatures those of others are fully disclosed; the collection of mathematical and physical truths is essential, on the mere chance that something will come of it—"il est toujours utile de penser même sur des sujets inutiles."[81] Then reviewing what science has already accomplished, Fontenelle marvels at the telescope and the air pump, at the "fécondité sans borne" of physics. He concludes that the numerous observations and experiments necessary for true advancement are impossible through individual effort and can be accomplished only through co-operation, such as is represented by the Academy.

Just what did the establishment of the Académie des Sciences do for the progress of science? Through the liberality of the royal treasury great laboratory facilities were established, especially a model observatory, the first worthy home for the study of astronomy, destined to be copied by other nations.

[80] Maindron, *op. cit.* (1881), p. 689, quoting *Le mercure galant*.

[81] Contrasting history and science, Fontenelle says: "History is but the spectacle of perpetual revolutions in human affairs; let us rather take an interest in the movements which control all the forces of nature."

THE ACADÉMIE DES SCIENCES

Leisure was secured to a number of gifted men to devote that time to experiment which otherwise would have had to be spent in gaining a livelihood, and an opportunity was given to certain great scientists, such as Huygens, Mariotte, and Pecquet, to reach other lesser minds and fertilize their thoughts and methods.

The work accomplished by joint experiments was undoubtedly of some significance, but, as a whole, the method of working together had proved a failure, and the greatest things were accomplished by individual discoveries. The fact that such men as Huygens withdrew shows disapproval of that method, and this is further emphasized by the enactment in 1699 of the statute discontinuing joint work. But the scientific method of exactness was highly advanced; astronomical instruments were immensely improved; previously such painstaking labor as Picard's had not been approached. The chemists had fostered love and interest in their experiments; the anatomists had won even from the geometricians sympathetic interest for their researches. The co-operation of the physicists with the physicians resulted, according to Daremberg, in the latter often adopting physical methods in their study of organisms, a proof of a beneficent exchange of ideas. The fact that the academical sessions were battle grounds of conflicting opinions can be no cause of regret, as these battles had to be fought, and in the relatively enlightened members of the society the new ideas found better champions than they might have found in the outside world.

Comparing this sketch of the Académie des Sciences with that of its sister-society in England, it must of course be borne in mind that, while through Newton's work the Royal Society had reached its zenith by 1700, the Academy was but in its infancy, and its greatest men fill the annals of its work during the eighteenth century, mainly 1750–1800.

One most important feature must not be lost sight of: Owing to the predominating political position of France on the Continent, and to the fact that she was, in the *siècle de Louis XIV*, the *Kulturträger* of the Continent, that her language and her manners were the model of the "correct," her Academy came to partake of the quality of "perfection" attached by the next century to things sprung from French soil; it became the model of many other academies—much as the court of Versailles was the model of other courts in the eighteenth century, and therefore potentially it must be given credit for a great deal that the next century was to accomplish by means of other learned societies.

CHAPTER VI

GERMAN SCIENTIFIC SOCIETIES

As we turn to the learned societies in Germany we must note some features which distinguish the situation there from that in Italy, England, and France.

Science seems somewhat more backward in Germany than in the other countries. Manual skill, the essential element in laboratory work, and instrument-making, had been more highly developed at Nürnberg than in other parts of Europe;[1] but the application of this skill along scientific lines had not yet taken place to any great extent; and it must be admitted that except for Kepler and Guericke, the greatest scientists throughout the seventeenth century belonged to other countries. Medical science was at a low ebb, and what knowledge there was came from Padua, Montpellier, and Leyden. Interested amateur scientists —other than alchemists—are less often met on German soil than in the other countries, due, perhaps, to the innate *Schwerfälligkeit* of the German, which is never compatible with the versatility of the amateur.

The backward condition of Germany is best illustrated by the low state of development of its vernacular. While in France and England great writers had written or were writing in the vernacular at this period, in Germany interest in the mother-tongue was just beginning and Latin was still without exception the language of the learned; so that in so far as the spread of scientific interest among the people depended upon the vernacular, it would necessarily be less rapid in Germany than in the other countries. This reacted upon the interest in science in still

[1] Joh. Neudoerffer, *Nachrichten von den vornehonsten Künstlern und Werkleuten so innerhalb 100 Jahren in Nürnberg gelebt haben 1546* (Fortsetzung von Gulden, 1660).

another way. Much of the enthusiasm which in other countries was bestowed upon the cultivation of experiment was given, in Germany, at this time to the cultivation and development of the German tongue. The Sprachgesellschaften, as, for instance, the Fruchtbringende Gesellschaft,[2] absorbed the very material from which the Fellows of the Royal Society were recruited. So that while in Italy and France the societies interested in language were strictly differentiated from those that were working in science, in Germany the one appeared as a preparatory stage for the other. These Sprachgesellschaften were at the height of their activity in the second half of the seventeenth century; and the man who stood out as champion of the learned societies in Germany, Leibniz, conceived of their work as an integral part of his scheme. Yet we shall see that in spite of these adverse factors, Germany plays a significant part in the history of the learned societies during the time we are considering.

Of the several German learned societies we shall mention first the Societas Ereunetica, a society founded in 1622 by Joachim Jungius, a most interesting personality; second, the Academia Naturae Curiosorum, a society of physicians only, recognized and supported by the imperial power; third, the Collegium Curiosum sive Experimentale, which was the creation of one man, and represented little more than the attempt of an enthusiastic teacher to gather students and friends for scientific work and discussion; fourth, the Berlin Academy, the only one which in time was destined to rank with the societies in France and England. As this society was founded in 1700, it falls practically outside of the present investigation; through the personality of its founder, Leibniz, however, it is closely connected with the seventeenth century; hence Leibniz, his multiform activities and his attitude toward learned societies, will be taken up, and it will become evident that through his efforts not only the foundation of the Berlin Academy, but also the origin of three other

[2] F. W. Barthold, *Geschichte der Fruchtbringenden Gesellschaft*.

societies, namely, those of Dresden, St. Petersburg, and Vienna, can be traced to the seventeenth century.

The Societas Ereunetica was established at Rostock in 1622. Its founder, Joachim Jungius (1587–1657), was a man whom Leibniz classes with Galileo and Pascal.[3] It was granted to him to understand the deficiencies of his own time, and to point out those ways of improvement which the future, after a laborious process of evolution, ultimately adopted. In his early years he belonged to an interesting group of pedagogic reformers, of whom Ratichius (Ratke) (1571–1635) was the leading spirit.[4] Their central idea was thus expressed by Jungius:

> It is the absolute truth that all arts and sciences, as, for example, the art of government, knowledge of weights and measures, of medicine, astronomy, architecture, fortification, can much more easily, comfortably, correctly, perfectly and explicitly be taught and promulgated in German than in Greek, Latin or Arabic.[5]

From enthusiasm for school reform Jungius turned to the study of natural sciences. He first studied pharmacy at Rostock,[6] afterward went to Padua and Pisa,[7] where he studied under Cesalpino. While Jungius was interested in physics, mineralogy, and zoölogy, his chief interest lay in the fields of botany and entomology.[8]

How did Jungius come to plan the formation of a learned society? One might infer that the reports of the Accademia dei Lincei suggested it; but Jungius' biographer points out that in a

[3] G. E. Guhrauer, *Joachim Jungius und sein Zeitalter*, p. 141.
[4] *Ibid.*, p. 26.
[5] *Ibid.*, p. 30. Cf. *ibid.*, pp. 36 ff. This emphasis on the vernacular, coupled with other reforms, attracted much attention in Germany. It influenced among others the famous reformer Comenius. Augsburg merchants wanted a school started along these lines, but Ratke seems to have lacked the executive ability necessary to organize such an enterprise.
[6] *Ibid.*, p. 45. [7] *Ibid.*, p. 50.
[8] He was the author of important botanical works, which were well known and published by the Royal Society (Sachs, *op. cit.*, p. 58).

book of Valentin Andreae, as early as 1619,[9] five years before the publication of the *New Atlantis*, a suggestion was made for founding an academy or college of natural sciences and the arts and crafts closely connected therewith. It was due, then, perhaps to Andreae's suggestion that Jungius, in 1622, founded his society.[10] He summoned disciples in the following manner: "Scholars in North Germany," he wrote, "have found a method of proving the falsity of Jesuitical teachings; they can show the only correct way of investigating natural phenomena, but inasmuch as this way is costly (on account of the obvious need of instruments) they need the help and support of all; especially of the rich, who are solicitous for the truth."[11] The purpose of the society, as stated, was to free all arts and sciences, which depend on reason and experiment, from sophistry, and reduce them to demonstration; to impart them through proper instruction and increase them through happy invention. Thus we have a society evidently designed to perform experiments. Secrecy was enjoined upon its members.[12] The motto of the Ereunetica was truly Baconian: *"Per inductionem et experimentum omnia."* The Academy was short-lived; no traces of it are found later than 1624. Its main interest lies in its early date and the personality of its founder,[13] of whom Goethe said: "If people had fol-

[9] The title of Andreae's work is *Reipublicae Christianapolitanae descripto*. Herder rediscovered Andreae, and considered him the Bacon of Germany.

[10] *Ibid.*, p. 69.

[11] Note the appeal to a lay public (Guhrauer, *op. cit.*, pp. 70 ff.).

[12] Among these were one physician, one senator, and one mathematician.

[13] Jungius later became professor of mathematics at Lübeck, and continued to advocate experimental methods. He preached experimental physics and mathematics as opposed to astrology (Guhrauer, *op. cit.*, pp. 79 ff.); similarly when later he became rector of the Johannaeum, the high school in Hamburg, he insisted on modern methods of teaching in contrast to the methods prevailing in universities. One of his pupils wrote: "If the students of the Gymnasium at Hamburg could and would take to heart the treasures heaped upon them, treasures for which people here have lost all appreciation and desire, there would come forth from Hamburg a gleam of science even as the Greeks came forth from the Trojan horse" (*ibid.*, p. 125).

lowed the advice of Jungius as to methods of study, the world would have reached, a hundred years earlier, that point at which it is today."[14]

The second learned society in Germany which we propose to notice is the Collegium Naturae Curiosorum. It antedates both the Cimento and the Royal Society, as it was founded in the year 1651, and it is still in existence today. It was not a society in the usual sense of the word, not a body holding joint meetings at which experiments were shown, or business transacted. It was merely a society of physicians which had its headquarters wherever its president happened to be located. Its main function was the publication of a scientific paper containing the original researches of its members and other investigations of importance to medicine. It is apparent that in type of membership, and in method, it differed fundamentally from the Cimento or the English or French societies, and resembled them only in a general way, as dealing with scientific questions, and in being officially recognized by the Emperor.

Its history is told in its publications[15] and again by Büchner (1755),[16] one of its presidents. Dr. Lorenz Bausch, *Stadtphysicus* of Schweinfurt, had studied medicine for two years in Italy and was much impressed with the work of the Accademia dei Lincei. He had a museum of "rarities," was learned in chemistry and botany, and had read Bacon's books. In the fall of 1651 he proposed[17] to the physicians of Schweinfurt to form an academy.

---

[14] *"Leben und Verdienste des Doctors Joachim Jungius,"* Goethe's *Nachgelassene Schriften*, published with Guhrauer's *op. cit.* (Appendix).

[15] Miscellanea curiosa.

[16] Andrea Elia Büchner, *Academiae Sacri Romani Imperii Leopoldino-Carolinae*. All documents are collected in Büchner, *op. cit.*, Vol. II: *Sacrae Caesareae Majestatis Mandato et Privilegio Leges*.

[17] *Ibid.*, pp. 19–24.

This plan was accepted and in January, 1652, a society was established, consisting of only four members.[18]

The society's aim was, in the words of the program, "the advancement of medicine and pharmacy through observation; by presenting observations in monographs, and communicating them to the members for correction and further elaboration."[19] The most important points of the plan of campaign were, that the president assign a topic to each member; that this topic be changed each term; that the truth of all statements be definitely demonstrated.

Little is known of the society's activities during its first ten years.[20] In 1661 Dr. Philipp Jacob Sachs von Lowenhaimb, the *Stadtphysicus* of Breslau, a man of information and prominence, joined them. This established a connection with an important commercial city, and the society gained in reputation so that in 1662 it counted twenty-five members.[21] It was proposed to publish a volume of its work in 1662,[22] but the difficulties seemed insurmountable and until 1670 only some monographs, mainly of Sachs and Bausch, appeared.

In 1670, through Sach's activity (Bausch died in 1665),[23] the society was enlarged and fundamentally reorganized. A letter[24] was sent broadcast to physicians inviting them to join, and specimen observations were published to indicate the scope of the work, "magnum opus aggredimur nec quae quoque viribus

[18] J. D. F. Neigebaur, *Geschichte der Kais. Leopoldina-Carolinischen Akademie der Naturforscher*, pp. 17 f.

[19] Büchner, *op. cit.*, pp. 19 ff.

[20] Leibniz often referred to its members, but criticized them for collecting from books rather than from observation (Foucher de Careil, *Œuvres de Leibniz publiées pour la première fois d'après les manuscrits originaux avec notes et introductions*, VII, 80).

[21] Büchner, *op. cit.*, p. 57.

[22] *Ibid.*, p. 414.

[23] *Ibid.*, p. 63.

[24] *Miscellanea curiosa*, Vol. I, *passim*.

GERMAN SCIENTIFIC SOCIETIES

hisce conveniant." The statutes[25] of the society were, as a whole, not changed. Membership was extended to "doctors, *licentiati* or those approximating them in learning,"[26] but the society remained eminently an association of German medical men.[27] Each member was required to choose a subject of investigation, or have it chosen for him. This topic was later printed after his name, and formed a most essential feature of his membership. The statutes directed that observations, experiments, inventions, problems, communications, should be collected into a volume with the name of the author, and this should be published yearly under the title *Ephemerides*. A director was appointed, in order that there might be a permanent officer to whom scientists could send their work. Soon the first volume of the *Miscellanea curiosa sive Ephemeridum medico physicorum Germanorum* was published.[28] In its Preface an appeal was made to physicians to devote themselves to science, for they lived in a time when even princes, the Medici, Louis XIV and Charles II, were actively interested. The *Philosophical Transactions* were to serve as the model of the *Miscellanea*, but they were to differ from the English publication in dealing only with matters which related to medicine, namely, physics, botany, anatomy, pathology, and chemistry. The editors explained the need of the undertaking they proposed on the ground that the spread of beneficial knowledge was slow and that unless some such method as a medical paper were devised it was impossible to reach the busy man. As a symbol of the society the ship "Argo" was adopted, the golden fleece signifying scientific truth.[29]

The first volume contained papers by Sachs and Bausch.

[25] For laws see *ibid.*, Vol. II. [26] Neigebaur, *op. cit.*, p. 17.

[27] One professor of eloquence had joined in 1677, and there were four foreign members by 1693.

[28] Neigebaur, *op. cit.*, p. 12. It was the work of Sachs and Lucas Schroekh, the first to hold the important office of director of the *Miscellanea*.

[29] *Miscellanea curiosa*, I, 2 ff.

Twelve of the twenty-seven contributions were on zoölogy.[30] There was much talk, and many pictures, of monstrosities, and abnormalities, in the spirit of the time. According to the *Miscellanea*, this first volume caused great satisfaction in the learned world; "cum curiosi academici viderent ipsorum novis inceptis conspirare eruditorum plurimorum consensum. . . . ."[31] It attracted at once the attention of the Royal Society, and Oldenburg wrote to the editors: "We do not doubt that Germany, ever fertile in learned men, will greatly add to the store of knowledge." He moreover expressed his gratification that the *Philosophical Transactions* had served as a model.[32]

As to the succeeding volumes, we notice that many contain translations of works from the vernacular into the Latin tongue.[33] There are articles by Leibniz, by Christopher Sturm on Borelli's *De motu animalium*,[34] etc. As to their general tone, there is a great deal of flattery of the Emperor in each Introduction. German sentimentalism on the death of the presidents is expressed in Latin verse or, as in the case of Fehr, in sixty pages of prose.[35] There is a lively tone of exultation in the frequent reviews of the work being accomplished. "Quot herbas decerpimus, quot min-

[30] Evidently great care had been spent upon these zoölogical articles, although Carus (*op. cit.*, p. 412) thinks that they were below the level of the contemporaneous publications of other countries.

[31] *Miscellanea curiosa*, Vol. II, Preface. Cardinal Leopold de Medici thanked the editors especially for the publication of Francisco Redi's work; the Jesuit experimenter, Athanasius Kircher, and the famous Danish anatomist, Thomas Bartholinus, expressed the unusual interest in the publication. The volume went into a second edition in 1684 (Neigebaur, *op. cit.*, p. 12).

[32] We read: "Dr. Croone in May, 1670, produced a printed paper published at Leipzig in imitation of the *Journal des sçavans*, and the *Philosophical Transactions*, viz., *Miscellanea curiosa*." Oldenburg, on February 16, 1671, gave an account of it (Birch, *History of the Royal Society*, II, 437 ff.). It was found that some observations contained in the work were considerable, and that its compilers were to be encouraged.

[33] Works of Grew (*Miscellanea curiosa*, VII, 330), Boyle, Helmont, Della Porta, Paracelsus (*ibid.*, XXV, 2).

[34] *Ibid.*, Vols. IX, X.   [35] *Ibid.*, Vol. XV.

eralia scrutamur, quot viventium inquirimus varietatem et viscera, tot hymnos Deo canimus." They add apologetically: "We seem to applaud ourselves because nature from its horn of plenty has poured the flood of its secrets upon us."[36]

How the attention of the Emperor was turned to the society is not clear. The bubonic plague, which then raged, had made the art of medicine seem even more important than before, and given prominence to the efforts of physicians who tried to widen its field of efficiency. By 1677 the Emperor accepted the rôle of patron, and thereafter the society was reorganized as Sacri Romani Imperatoris Academia Naturae Curiosorum. The statutes remained almost unchanged but the new significant motto *"Numquam otiosus"* was adopted. From now on it grew steadily in importance and in numbers.[37] Its later presidents came from Nürnberg and Augsburg, bringing it into essential contact with large commercial cities. By 1687 the Academy adopted the name Academia Caesarea Leopoldina, and its privileges were still further extended by the Emperor.[38] It had, for example, officially the same rank as the University of Vienna, full license to print, and copyright privileges.[39]

But the society was confronted with many obstacles. First of all there was financial embarrassment[40] and very little aid from the Emperor. The imperial protection, moreover, which appeared so promising did not count for much in this time of declining imperial power. The workers were located at widely distant places and had to depend on written communications. The lay public could not be interested in publications as highly specialized as the *Miscellanea;* in fact, Prutz well says that "the

[36] *Ibid.,* Vol. VIII, Preface.

[37] *Ibid.,* Vol. XIII. By 1683 the membership was 159.

[38] The president and director of the *Miscellanea* were raised to the dignity of count palatinate, with the right of legitimizing illegal children, sanctioning adoptions, etc. (Neigebaur, *op. cit.,* p. 3).

[39] *Ibid.,* p. 11. [40] *Miscellanea curiosa,* Vol. VIII, Preface.

paper might have gone on a hundred years, without the general reading public learning of its contents, or even of its existence."[41] Then there was continual discord and jealousy among the members. Yet in spite of all these difficulties the society persisted. Undoubtedly the *Miscellanea naturae curiosorum* take a high rank among the publications of the end of the seventeenth century; they are referred to in the *Philosophic Transactions,* in the *Journal des Sçavans,* and in Bayle's *Nouvelles de la république des lettres.*[42] Bayle praised the *Mélanges curieux*[43] as showing the indefatigable industry, invention, and genius of the Germans; and pointed out that this publication was really a rival of the academies.

Several features of the work of this Academia Naturae Curiosorum must be commended. In comparison with the English and French societies, it represented a tendency toward greater specialization. It maintained that medical science must necessarily rest upon a basis of scientific experiment. In insisting upon original work by each member it created a body of men seriously engaged in experimental work. It adopted one of the most essential features of other learned societies—that of digesting and spreading the new knowledge which was rapidly accumulated. Through its publication, it brought the news of progress in medical science to many a physician whom it would not have reached. All this it did, apparently, with a minimum of organization and official machinery.

But it in no way created opportunities and means of original work, and in the last analysis it seemed bent on collecting the work of men who had scientific interests rather than on creating such interest or calling it forth. In short, it admirably performed

[41] R. Prutz, *Geschichte des deutschen Journalismus,* p. 275.
[42] Pierre Bayle, *Nouvelles de la république des lettres* (Oct., 1685), p. 389.
[43] Bayle's translation of their name.

the function of editor of a learned paper, but, except in name, hardly deserves to be classed as a learned society.

The next scientific society which presents itself in Germany, the Collegium Curiosum sive Experimentale, was entirely different from those previously considered. At the University of Altdorf, which stands out as the one progressive German university,[44] Christopher Sturm held for thirty-four years the position of professor of mathematics and physics. He had been present at sessions of the Cimento,[45] and had gained a knowledge of their work and methods; he had studied in Holland and England, kept up a correspondence with learned men, for example with Halley and Hack,[46] and wrote treatises on the most varied subjects.[47] But his chief claim to fame was that he was considered the most skilful experimenter in Germany at the end of the seventeenth century.

He had in his home a private physical laboratory equipped with all necessary instruments, which he used in his university lectures. In 1672 he decided to form, among the devotees of experimental science, mainly his students, the Collegium Curiosum sive Experimentale, in imitation of the Accademia del Cimento. The invitation issued for this purpose strikes the keynote of the enterprise:

> Everybody realizes how much more this century knows of nature's laws than the preceding, and it is well known that the experimental method has brought this about. The Collegium Curiosum is established with the express purpose of doing wonderful experiments with the barometer, telescope, camera obscura, pendulum, thermometer, hygrometer (though some of this may seem only boy's play). Not "auditors" but "spectators" are wanted *quod modernae philosophiae proprium est ad experimenta ipsi oculorum sensui exposita.* . . . .

[44] See below, chap. viii.   [45] *Atti et memorie,* I, 305.
[46] Birch, *History of the Royal Society,* IV, 383.
[47] Sturm wrote textbooks on arithmetic which were long used throughout Germany, a book on planets, articles on Borelli's *De motu animalium,* etc., etc.

Sturm promised to show publicly what he had at times demonstrated to his students, and had for this purpose collected all the necessary instruments, especially air pumps, to enable the society to repeat the experiments of Schott, Boyle, and Guericke, etc.[48]

The original members numbered twenty—none of whose names posterity has made famous, none indeed who are called doctors, hence I assume that most were students—but their number increased steadily.[49]

The account of the work of this society has come down in Latin in two folio volumes entitled *Collegium Curiosum sive Experimentale,* Altdorf, 1672, published in 1676 and 1683. The first volume consists of accounts of sixteen experiments, some, according to the announcement, first made by the Collegium, some merely repeated by it. These sixteen *experimenti tentamina*[50] represent the highest type of the physical research of the time.[51]

The second volume comprised fifteen experiments[52] made

[48] *Collegium Curiosum sive Experimentale,* Preface.

[49] Twenty-one new members in 1675; thirteen in 1679; eighteen in 1681; twenty-seven in 1683; fifteen in 1684.

[50] 1. Diving bells. 2. Camera obscura—explanation of eye. 3. The Barometer experiment (the same as in the *Saggi*). 4. Syphons. 5. Hydrostatic experiments. 6. Water pump. 7. Experiment to show the atmospheric pressure with a tube of water 36 feet high. 8. Capillarity. 9. Thermometer. A differential thermometer of importance in meteorology. 10. Airship invented by Franciscus Lana. 11. Mechanics. Laws of lever and simple machines; an apparatus to show the law of the parallelogram of forces. 12. Universal language. 13. Air pump: Boyle's and Guericke's and their experiments; the society made improvement on the pump. 14. (*a*) Hygroscope. (*b*) Chronometer. 15. Microscope. Telescope. 16. Lenses. Refraction.

[51] Rosenberger, *op. cit.,* II, 209.

[52] The fifteen experiments related to: 1. Diving bells. 2. Air pumps. 3. Barometer. 4. Hygrometer (toys). 5. Syphons (Arcadian well). 6. Glass *Tränen*. 7. Lever. 8. Tubes, trumpets (ear trumpets). 9. New thermometer experiments. 10. The Magdeburg hemispheres. 11. Mechanics of muscles. 12. Von Helmont's experiment. 13. Hygroscope. 14. A perfection of Guericke's pump.

from 1685 to 1692, in great part a continuation of the previous ones "as wonderful as those before."[53]

The importance of the society was that it spread interest in science and trained clever experimenters; indeed, the two volumes it published were in later years used as a textbook of experimental physics. Besides, this society is of historic interest because it closely resembles the milieu in which experimental science is fostered today. Our modern institutes connected with universities, drawing aid and inspiration from the university professors, resemble greatly this Collegium Curiosum sive Experimentale at the home of the professor of physics at Altdorf.[54]

In studying the foundation of the Berlin Academy, the only German learned society which parallels the Royal Society, or the Académie des Sciences, a most important point of difference must be noted. The beginnings of the other societies have been traced to enthusiastic amateurs, who met informally to enjoy experimentation, and who established themselves, ultimately under royal patronage, into definite corporate bodies. Such was not the beginning of the Berlin Academy. Its existence was due to the perseverance and farsightedness of one man, Gottfried Wilhelm von Leibniz; and a study of this man, his attitude toward learned societies in general, and his efforts to found the society in Berlin form the only possible introduction to an understanding of the establishment of the Berlin Academy. Moreover—as was said above—since the Berlin Academy was not founded

[53] Besides these experiments in physics many things were undoubtedly tried, so, for instance, Haeser mentioned that at Altdorf under Sturm's guidance work was done on the transfusion of blood (Haeser, *op. cit.*, II, 421).

[54] The important question whether the society outlived its founder, I cannot answer. The *Encyclopaedia Britannica* asserts that it lasted some time after Sturm's death. Sturm planned another larger society, and sent an *Epistola invitans ad observationes magneticae variationis communi studio junctisque laboribus instituendos*—an unsuccessful plan for a magnetic "world-union" such as was accomplished one hundred years later by Humboldt.

until 1700, its history properly lies outside the scope of this essay, except in so far as Leibniz and his rôle are concerned.

The importance of Leibniz as a *Kulturträger* has been fully recognized; Frederic the Great called him "a whole academy in himself." Keller aptly says:

> Among those who incarnated the change of civilization, no name next to Bacon's should be given greater prominence than that of Leibniz. Nor is it enough to call him a pioneer; he is the creator and spokesman of those special elements which characterize modern times in distinction to the Middle Ages.[55]

"Nobody," says Harnack, "excelled him in concentrating in himself all the forces of the century."[56] "He was," in Kuno Fischer's words, "like his own monad: *chargé du passé et gros de l'avenir.*"[57] Indeed, if we leave aside his activity as a philosophical writer, which to the less instructed seems his main activity; if we exclude his activities as diplomat and international statesman, which undoubtedly seemed to him of prime interest; if we exclude his work in jurisprudence, which was his profession and omit his marvelous historical research, which entitles him to be regarded as a forerunner of the modern critical editor of sources, and of which Harnack says, that it alone would have been a life work for most men; if we neglect his tireless activity in reconciling the Catholic and Protestant faiths, which was his lifelong aim and cherished hope, we find in the residue of his activities an active participation and co-operation in all those great movements which served to build up a new era.[58] For he was an opponent of the accepted university education; he aimed at the reconstruction of the entire educational fabric, always empha-

---

[55] Dr. Ludwig Keller, *Gottfried Wilhelm Leibniz und die deutschen Societäten des 17 Jahrhunderts*, p. 2.

[56] Adolph Harnack, *Geschichte der Königlich Preussichen Academie der Wissenschaften zu Berlin*, I, 9.

[57] Fischer, *op. cit.*, p. 11.

[58] One exception must be noted—Leibniz' opposition to Toland's deism (Harnack, *op. cit.*, p. 9).

sizing the superiority of the vernacular, of *realia*, over book knowledge. A great mathematician, he loved mechanics and physics, so that as an old man he said to Peter the Great that he preferred making scientific discoveries to honor and fortune, though chance had made him pay more attention to history and politics. He was connected with the laboratory work of the Rosecrucians, was a pioneer geologist, a member of both the English and French societies and founder, on their model, of the third great academy of Europe. Moreover, he was a constant contributor to scientific papers such as the *Acta eruditorum* and the *Journal des Sçavans*, and established an important German periodical, *Monatliche Auszüge neuer Bücher*.[59] He found time to correspond with some of his greatest contemporaries, for example, Huygens and Spinoza.[60] He was cosmopolitan,[61] almost ubiquitous, in the realms of thought. He lived and thought in that larger world of our modern colonial interest. China, Ceylon, and the Pacific Coast were realities to him. So thoroughly modern does Leibniz seem to Harnack that he suggests, if Leibniz came to life today, he would only see what he definitely foresaw, even the revolutions in the technical arts coming within the compass of his prophetic imagination.

It is Leibniz, the reformer and critic of existing educational conditions, and the founder of the Berlin Academy, upon whom our attention will be concentrated.

To understand Leibniz' attitude toward education a few words about his *Lehrjahre* are necessary. He had obtained an orthodox Aristotelian education *chargé du passé*, but he had been years ahead of boys of his age. He had studied law at the most conservative university, Leipzig, and there read philosophy

---

[59] Wegele, *Geschichte der deutschen Historiographie seit dem Auftreten des Humanismus*, p. 652.

[60] There still exist 15,000 of his letters.

[61] At one time he was president of the Berlin Society; historian at Hanover; he longed to live in London and to become the historian of England; kept up his relations with the Académie des Sciences and wanted to direct the intellectual work of Austria and Russia (Harnack, *op. cit.*, I, 182).

with Jacob Thomasius, father of the reformer Christian Thomasius. Soon, he tells us, he became acquainted with the new school of thought.

It well chanced that the plans of the great Bacon about "the Advancement of Knowledge," and the deep thoughts of Cardanus and Campanella, and the evidences of a better philosophy as contained in the writings of Kepler, Galileo and Descartes, came into the hand of the youth.[62]

Later he studied mathematics for some semesters at Jena under Weigel.[63] Weigel was a great lover of mathematics and a skilful mechanician,[64] also an astrologer, astronomer, jurist and philosopher, an enemy of scholasticism, and a friend of the new thought. In 1673 he had established a mathematical society, Societas Pythagorea; he was interested in a *Tugendschule,* a new type of school along Comenius' lines. His lectures were so popular that he could find no room spacious enough for those who wished to hear him.[65] It is very evident that Weigel exerted a deep and lasting influence upon Leibniz.

Returning from Weigel's lectures to Leipzig, Leibniz distinguished himself in legal work, but on account of his youth failed to obtain his Doctor's degree at the first trial.[66] To this may be traced in part his hostile attitude toward universities, or rather this may have gone far to open his eyes to their overconservative features. He left Leipzig and went to the most progressive University of Altdorf where he took his degree, but declined an offer of a professorship. While there, he became a member—indeed secretary—of the Rosecrusian alchemist order, which was evidently an indication of his growing interest in experimental science.[67]

---

[62] Fischer, *op. cit.,* p. 40.   [63] *Ibid.,* p. 42.

[64] Dr. Bartholomei, *Zeitschrift für Math. und Phys.,* XIII, *Suppl.,* pp. 1–14; *Erhard Weigel,* p. 33. Weigel was the author of thirty inventions.

[65] Cantor, *op. cit.,* III, 36.

[66] Fischer, *op. cit.,* p. 46.

[67] Hermann Kopp, *Die Alchemie in älterer und neuerer Zeit,* p. 623.

These were Leibniz' *Lehrjahre*. His aversion to the educational system as it then existed comprised in its criticism school and university methods alike. Thus, he wrote:

> The teaching of youth should be centered not so much upon poetry, logic and scholastic philosophy as upon *realia*, history, mathematics, geography, *vera physica;* instruction in *realia* should be pursued in collections of rarities, the study of man in anatomical theatres, chemistry in the apothecary's shop, botany in botanical, zoölogy in zoölogical gardens. The pupil should forever move in the *theatrum naturae et artis*, receiving living knowledge and impressions.[68]

He opposed absolutely the emphasis which school and university placed upon Latin, because he wanted education to reach everybody, and thought that then, and only then, a better condition of society could be attained.[69] He also felt that too great a preoccupation with Latin might be injurious to mental development;[70] and lastly, he objected to Latin because to him it was allied with old thought, while the vernacular was the mouthpiece of the new times. It was, he thought, because Bacon and Descartes had written in English and French that scholasticism was dead in England and France; because German was not used in Germany that she still suffered from the bane of scholasticism.[71] He writes:

> Our learned men have shown little desire to protect the German tongue, some because they really thought that wisdom could only be clothed in Latin and Greek; others because they feared the world would discover their ignorance, at present hidden under a mask of big words. Really learned people need not fear this, for the more their wisdom and science come among people, the more witnesses of their excellence they will have. .... On account of the disregard of the mother tongue, learned people have concerned themselves with things of no use, and have written merely for the

[68] Foucher de Careil, *op. cit.*, VII, 52 ff.

[69] Keller, *op. cit.*, p. 4.

[70] Leibniz said, for instance: "We wrongly dull the intellects of people by learning Latin" (quoted in Harnack, *op. cit.*, p. 31).

[71] Fischer, *op. cit.*, p. 66.

bookshelf; the nation has been kept from knowledge. A well-developed vernacular, like highly-polished glass, enhances the acuteness of the mind and gives the intellect transparent clearness.[72]

Just as he objected to the use of Latin, he objected to the entire university system.[73] He called the universities "monkish," and accused them of being absorbed in trifles. They gave token of learnedness, never of their own judgment, and did not dare to advance anything for which there was no ancient authority. He praised in contradistinction to learned subtilty the lowly arts of the artisan and "practical man." In a book on carpentry he asserted he found more usefulness and truth than in learned works.[74] Again he wrote, "If I had to make a library, I should have only books of invention, experiment or historical documents,"[75] for in "Bücher so am wenigsten geistreich sind, sich immer ein oder ander guter Gedanke findet."

From the close of his student days (1667) to his death (1716) Leibniz lived away from university influences, in the close vicinity of courts and in intimate intercourse with noble personages. During these years two ideas never left him: one, the reconciliation of Protestantism and Catholicism; and the other—prosecuted with almost religious ardor—the founding of learned societies. He would forward progress, change, *Aufklärung*. Other men might have sought to trumpet forth their views by writing books or by teaching, but Leibniz had little faith in books and in existing channels of instruction. He felt

[72] Keller, *op. cit.*, p. 9, quoting from Leibniz: "Unvorgreifliche Gedanken betreffend Ausuebung und Verbesserung der Teutschen Sprache. Ermahnung an die Teutschen ihren Verstand und Sprache besser zu üben, nebst Vorschlag einer teutsch gesinnten Gesellschaft." He felt that the Fruchtbringende Gesellschaft had right aims, but failed because it strove to introduce German into poetry only and not into science.

[73] Friedrich Paulsen, *Geschichte des gelehrten Unterrichts auf deutschen Universitäten*, I, 347.

[74] O. Klopp, *Die Werke von Leibniz*, III, 229.

[75] Keller, *op. cit.*, p. 5.

that only by co-operation with men who sympathized with and shared his attitude could his ideas ultimately prevail.[76] A learned society, such as the Royal Society or Académie des Sciences, in contact with the living world, most appealed to him as the mouthpiece of his views; and such a society he desired to found on German soil. For us this persistence is of fundamental importance, for it is the best proof that to one of the most intelligent and most widely informed men of the second half of the seventeenth century the learned society seemed the best and only instrument to improve existing conditions. These attempts of Leibnitz will therefore be followed in some detail.

The first project for founding a learned society was made by Leibniz at the age of twenty-one. A proposal to found a semiannual journal, *Semestralia,* to review all publications[77] grew in Leibniz' mind into a plan for a Societas Eruditorum Germaniae.[78] It was to consist of a fixed number of learned men, was not to mingle in matters of religion, but to keep up a wide correspondence, to collect a universal library, to co-operate with the French, English, and Italian academies, to perfect medical science, to watch mathematical experiments, to collect experiments, and have general oversight over commerce and manufacture. Besides, the society was to have the right of granting licenses for the publication of books. Leibniz' plan was that every author should be required to indicate what matters, either new or useful to the state, his book contained, in order thus to limit the prevailing *scribicitas multorum*. As the Emperor refused the privilege of censorship, nothing further came of this plan.

Leibniz' next two proposals (1669-70) are most interesting. The first document bears the title *Grundriss eines Bedenck-*

[76] Harnack, *op. cit.,* I, 21.

[77] Foucher de Careil, "Bemühungen um ein kaiserliches Privileg für Plan seiner Semestria," *op. cit.,* VII, 1 ff.

[78] *Ibid.,* pp. 155 f.

ens von Aufrichtung einer Societät in Teutschland zum Aufnehmen der Künste und Wissenschaften.[79] It may be summarized as follows: Great inventions have been made in mechanics; great discoveries in physiology, chemistry, are kindling a new light; but everything in practical life is as before. The inventions have not yet been utilized to increase the comfort and happiness of humanity. In order to do this the Germans must found an academy as their neighbors have done. This will awaken the spirit of co-operation and correspondence of experienced men; inventions and experiments, now often lost to the world either because they are not communicated, or because the inventors lack the means of continuing their work, will be utilized.[80]

Leibniz then outlined his plan of the society. It was to be all embracing, comprising in its scope science, history, art, trade, commerce, police, medicine, archives, schools, machines, etc. For instance, dissection of the bodies of both men and animals should be carried on, on a large scale. Not pathology but plain physiology, the normal conditions of fermentation and chemical reaction, should be studied.[81] Similar to the work of our dispensaries, it was suggested that poor people be treated as clinical material. Manufactures were to be improved by means of new appliances. New suggestions were to be tried; manual labor and commerce to be improved. Further, Leibniz proposed, that education should be directed by the Academy, and orphans and foundlings should be educated along technical lines. Members of this society should teach *realia;* they should become traveling teachers—a novel type of missionary—and induce the Germans to refrain from studying abroad. The influence of the society should be exerted to give to nobles and clergy "the appetite for curiosity," or if they had it, help them to cultivate it. A journal should be founded to encourage correspondence, "that there be trade and commerce in sciences." It should collect useful

[79] *Ibid.,* pp. 27 ff.
[80] *Ibid.,* p. 48, par. 24.   [81] *Ibid.,* p. 53, par. 24.

thoughts, inventions, experiments, so often hidden in obscure places and among humble laborers. Surely—Leibniz suggests —if such people had a center where they could report their ideas, they would willingly do so. Furthermore, Leibniz added suggestions for self-supporting poorhouses and prisons. In conclusion, he emphasized the fact that his proposal was so much broader than the work of the Royal Society and Académie des Sciences that it should bring better results.

The second program[82] was published under almost the same title, *Bedencken von Aufrichtung einer Academie oder Societät in Teutschland zum aufnehmen der Künste und Wissenschaften*. Here we find as follows: the Germans were the first to make inventions, but are the last to utilize them. They did great work in mining, chemistry, mechanics;[83] all automatically moving things were invented by them (in Nürnberg and Augsburg); Regiomontanus, Copernicus, Tycho, and Kepler were Germans;[84] *medicina practica*, alchemy, and the art of the apothecary flourish in Germany; a German first tried the transfusion of blood. But inventions are exported from Germany and later, somewhat changed, imported as new from foreign countries.[85] The reason for this is that the Germans lack associations such as the Royal Society or Académie des Sciences. But this is not their fault. By the establishment of the Collegium Naturae Curiosorum and the Fruchtbringende Gesellschaft, the German nation has shown that it, too, could soar if it only were assisted.[86] German princes ought to follow the Royal Society, where the King, the Duke of York, Prince Rupert, and many nobles contribute to the expense of scientific work. They have their ministers communicate new and rare matter to the society; they admonish the directors of

[82] *Ibid.*, pp. 64 ff.
[83] *Ibid.*, pp. 65 ff.
[84] Incidentally he claims the invention of the telescope for his nation.
[85] *Ibid.*, pp. 66 ff.
[86] *Ibid.*, p. 80, par 16.

colonies and ship captains to bring whatever new they find to this *aerarium eruditionis solidae publicae;* they have the society formulate *interrogatoria,* instructions, *directoria* for travelers, ambassadors, miners, medical men, and artisans, to get deep into the mines of nature.[87] How rich is Germany in erudite men compared with England! But her skilful mechanics are starving and emigrating,[88] though exportation of mind is the very worst contraband.[89] The many alchymists would prove smart fellows if properly employed. Finally, an appeal is made to the nobles, on the plea of their health, which, according to Leibniz, will never be properly cared for without a society which encourages the application of the new discoveries of theoretical medicine to practice.

These two sketches contain Leibniz' program for learned societies, adhered to with only slight variation and additions throughout his life.

Indeed, his own free time was devoted to such a wide range of pursuits that he veritably seemed "an academy all by himself." Before leaving Mainz he was able to announce an imposing list of discoveries he had made, and plans for further discoveries in mathematics, mechanics, optics, hydrostatics, pneumatics, nautical science, chief among which was that of a calculating machine. These were the activities which brought him near to the Royal Society and Académie des Sciences in spirit. Soon he was to have the opportunity of coming into personal contact with these societies. Sent on a political mission to Paris, he stayed there from 1670 to 1676. Here he changed from the German to the European scholar.[90] He was intimate with many

[87] *Ibid.,* par. 17. See above, chap. iv.

[88] Foucher de Careil, *op. cit.,* Vol. VII, p. 84, par. 18: "Welche gescheid seyn, gehen fort und lassen Teutschland mit samt seiner Betteley im Stich."

[89] "Ingenia sind mehr vor Contrebande zu achten als Gold, Eisen, Waffen."

[90] Fischer, *op. cit.,* pp. 102 ff.

members of the Académie des Sciences,[91] and though he did not succeed then in obtaining an appointment to the French society, he undoubtedly had ample opportunity to follow its work. With the Royal Society he was at first in correspondence; later he brought his work—*Hypothesis physica nova*,[92] a new cosmic theory—and his calculating machine to London, and on his own application became a Fellow of the society (1673).[93]

The result of the experiences of these years is a new scheme, "Consultatio de naturae cognitione ad vitae usus promovenda instituendaque in eam rem Societate germana, quae scientias artesque maxime utiles vitae nostra lingua describat patriaeque honorem vindicet,"[94] a call to learned men to discuss the founding of an academy.[95] Leibniz' suggestion was that in Germany an association of men, imbued with a love of study, be established; free from financial cares and supplied with instruments for research, they would accomplish more in ten years than all humanity has done in centuries. The Germans should at last give scope to their skilled men, to their chemists and mechanicians, and to their language. They should obtain information, not *ex chartis*, but *ex naturae volumine et mentium thesauro*.[96] Important ideas can be obtained only from authors, whether learned or not, who for themselves investigate and experiment.[97] Observations must be collected, a German nomenclature established, a survey of existing problems must be made, separating those simple and already solved from the more difficult. Such

---

[91] Huygens introduced him to the study of higher mathematics. It was in these years that Leibniz developed his theory of infinitesimal calculus that brought him later into a violent quarrel with the Royal Society, which championed Newton's priority in the discovery.

[92] Fischer, *op. cit.*, p. 70: "Neque Tychonicis neque Copernicianis aspernando."

[93] Birch, *History of the Royal Society*, II, 475.

[94] Foucher de Careil, *op. cit.*, VII, 105.

[95] *Ibid.*, p. 101: "De fundatione scientiam provehendam instituenda."

[96] *Ibid.*, p. 105.   [97] *Ibid.*, p. 112.

systematic proceedings would put the Germans ahead of all other nations. Even the English and French societies have not accomplished so much, as they might have, had they more consistently heeded the useful. Germany should at last employ the German tongue, and follow the other nations who have dropped Latin, and thus opened arts and sciences to everybody, even to women and young people. An appeal is made to the members of the Fruchtbringende Gesellschaft and of the Collegium Naturae Curiosorum to unite in an imperial society. Leibniz even suggested forty-eight names[98] as the nucleus of such an association. The purpose of the society was outlined as follows: to find the true causes of physical phenomena; to make experiments and co-ordinate them; to investigate such matters only as might ultimately be of use in life; to employ the analytical and synthetical method in all experiments; to create an encyclopedia of all human sciences—for so great was the interdependence of the various sciences that one without the other could accomplish nothing. Every member was to assume a task and report upon it within a specified time and in intelligible language. All experiments were to be collected; all reports to be made by the experimenter himself.

Upon Leibniz' return from Paris he soon saw how impossible the scheme of an imperial society was, in view of the decentralized condition of Germany.

It was then that he entered the position that he was destined to hold until his death—librarian and confidant at the court of Hanover. During the next years we find scattered suggestions of societies. He planned a Societas Theophilorum vel Amoris Divini,[99] which was to supplement the work of the Jesuits, and cultivate those studies the Jesuits neglected, namely, natural

[98] Harnack, *op. cit.*, p. 31. Among these were Weigel, Helmont, Hevelius, Steno, Swammerdam, Leeuwenhoek, Tschirnhausen, Guericke, Homberg, Bartolini.

[99] Keller, *op. cit.*, pp. 8–9; Foucher de Careil, *op. cit.*, VII, 100.

science and medicine, and should teach, besides religion, *chymica et arcana naturae*. In 1681 Leibniz planned a magneto-mathematical society.[100] Then historical study seemed for a while to monopolize his mind. In Frankfurt he discussed with Hiob Ludolf the plan of an imperial German historical society,[101] similar in scope to the society which, many years later, created the Monumenta Germaniae.

During the years 1687–90 Leibniz undertook a journey through Germany and Italy to collect materials for his history; while in Rome he became a member of a physico-mathematical society,[102] and there he conceived the novel idea that Italian cloisters should be devoted to experimental study and become branch academies.[103]

With the marriage of the daughter of the Duke of Hanover to the Elector (later King), Frederick I of Prussia, Leibniz' previous indefinite ideas regarding the foundation of a German learned society were gradually focused into the plan of creating such a body in Berlin. When he heard that there were regular meetings of scientists, allotment of problems, etc., at the house of the diplomat, Spanheim, he addressed a series of letters to him, hinting at founding a "Societas Electoralis Brandenburgica exemplo Regiarum Londinensis et Parisiensis,"[104] for he believed now that the ruler of Prussia was the proper person to head such an enterprise. The type of undertaking which he outlined did not differ widely from those previously sketched. Leibniz seems surer than ever of the insufficiency of the Royal Society and of the Cimento and the Collegium Naturae Curiosorum. The Berlin society was to be but the central station of a number

---

[100] Harnack, *op. cit.*, I, 35.

[101] Wegle, *op. cit.*, p. 598; Harnack, *loc. cit.*

[102] Fischer, *op. cit.*, p. 201.   [103] *Ibid.*, p. 12.

[104] Foucher de Careil, *op. cit.*, VII, 599 ff.: "Denkschrift über die Errichtung einer Chürfurstlichen Societät der Wissenschaften"; Harnack, *op. cit.*, II, 35-42.

of branches. The objects of these societies were to be eminently practical, as, for example, the encouragement of agriculture, the draining of marshes, the discovery of mines, provisions for public health, guarding against epidemics, the education of youth, etc. As regards science, Leibniz stated his aim as the welding of theory and practice.

Not through Spanheim, but through the Electress, Leibniz' plans of founding a scientific society came near to accomplishment. Perhaps at the suggestion of Jablonski, the court preacher, she decided to erect an observatory in Berlin, modeled on that at Paris. The plan was reported to Leibniz, who was charmed with it, but added: "Cela vous pourra engager cependant à aller plus loin et penser encore à d'autres sciences curieuses."[105]

The founding of the Berlin Academy was, however, destined to be an outcome not of this, but of a plan of calendar reform[106] which at the time occupied Protestant Germany.[107] When in 1699 the *Corpus evangelicorum* adopted the Catholic calendar, it was necessary to establish a commission in Brandenburg, to supervise this change. Inspired by a previous suggestion of Weigel's,[108] Leibniz proposed that the Elector should keep the monopoly of calendars, and from funds thus accruing establish an observatory and a learned society. This plan was submitted to the Elector,[109] and its acceptance was the beginning of the Berlin Academy.

[105] Harnack, *op. cit.*, I, 47.

[106] I.e., leaving out eleven days as per the Gregorian calendar (*ibid.*, II, 58).

[107] *Ibid.*, I, 64 ff.

[108] Weigel, the professor of mathematics in Jena who was mentioned before as a mind kindred to Leibniz, proposed in 1694 to the Diet that an imperial Collegium Artis Consultorium of twenty men be established, which was to have a monopoly of calendar reform, and that the funds thus attained should go to the society for work in astronomy, mathematics, and arts (Bartholomei, *op. cit.*, p. 33).

[109] Harnack, *op. cit.*, II, 58 f.: "Jablonski Untertaenigster Vorschlag wegen Anrichtung eines Observatorii und Academiae Scienciarum."

GERMAN SCIENTIFIC SOCIETIES

"Seldom," says Harnack, "has an undertaking been started with so carefully elaborated a program,"[110] for Leibniz' previous speculations on societies were but preparatory to their realization in the Berlin Academy. Jablonski, entirely in Leibniz' spirit, made definite suggestions. He proposed to erect an observatory with a complete college of science including physics, chemistry, astronomy, geography, mechanics, optics, algebra, geometry, etc., because an opportunity had fortunately presented itself to do so without expense.[109] The rules of the Royal Society and Académie des Sciences were to be copied and improved. The president was to be Leibniz, whose great *eruditio in omni scibili*, also *stupenda inventa promotae matheseos*, eminently fitted him for the office. Kirch, a pupil of Weigel, the leading German astronomer, was to be put in charge of the observatory.[111] Several savants, for example, Tschirnhausen and Sturm, were mentioned as correspondents. The plan was to build over the middle wing of the royal stable an observatory, an assembly-room, a library, a room for instruments, and the apartment of the astronomer. Instruments would be easily acquired. All was to be established from the money hoped for from the calendar monopoly (2,510 *Thaler*).[112]

In accepting these proposals, the Elector made the significant suggestion that the cultivation of the German language should be added to the program of the society. Thus he was the author of its philological and historical features, in which regard it differs from the London and Paris body.[113] Leibniz made several characteristic additions to Jablonski's plan.[114] He did not wish the observatory to be the main feature, nor did he wish the society to stand for *curiosa*, as did the Paris, London,

[110] *Ibid.*, I, 73.

[111] *Ibid.*, p. 114.

[112] *Ibid.*, pp. 74 f. In the tables of forecast of expense 200 *Thaler* were allowed for instruments, 100 for printing, 500 for Leibniz, 500 for Kirch.

[113] *Ibid.*, p. 78.   [114] *Ibid.*, II, 72 (Leibniz' letter to Jablonski).

and Florence societies, but for *utilia* such as interest in agriculture, manufacture, commerce and food. A laboratory was to be established at once.[115] All appointees of the state were to correspond with the society. He expressed the hope that German nobles, like the English nobility, might develop scientific interest, and that scholars and university people should become affiliated with the society. The secretary was to be a young physician who would understand mathematics, mechanics, and chemistry.[116]

Soon, on July 11, 1700, the charter was obtained,[117] and the society with its statutes was constituted. According to the example of the Royal Society, there was to be a governing council which appointed the Fellows of the society. Meetings were to be held for three subjects: *Res physico-mathematicae, Lingua Germanica, Res literariae*. The *Acta* of the society were to be published as *Diarium eruditorium*. The society was to have an observatory, laboratory, library, museum, "rarities," *theatrum naturae et artis*, animals, and plants.[118]

But all this was on paper, and it was Leibniz' task to get the enterprise started. How busily engaged he was with this task is shown by a list written in August, 1700, of sixty-three matters to be attended to, where to meet, what to publish, whence to get instruments, with whom to start correspondence, etc. For, as we saw, he had waited for this opportunity a lifetime and was eager to see it improved to the utmost. It required, however, ten years before the society was in any sense established;[119] before

[115] "Recht gute Pendula sind hochnoethig ad mensuram temporis. Die Gerickesche Instrumenta werden wohl a propos kommen. Barometra, thermometra et hygrometra sind auch nötig, etc."

[116] A provision not carried out in the appointment of Jablonski to this post.

[117] For the historical student there is interest in the clause in which the history of Brandenburg, not only the cultivation of the language, but German and especially both ecclesiastic and profane, is recommended.

[118] Harnack, *op. cit.*, I, 92 ff.

[119] *Ibid.*, pp. 173 ff.; also II, 205 ff. The formal opening exercises of the society were held January 19, 1711, in the meeting hall of the Observatory.

it met at its own quarters, published its work, the *Miscellanea*, and obtained its new statutes. And after this a period of decline followed,[120] and Leibniz died feeling that the society was all but extinct.

In the first years Leibniz, though residing in Hanover, was very active; but there seems to have been disappointment very soon. The eighty Fellows who were appointed did not come up to his standard.[121] There were continual financial difficulties, for it was understood that the society was not to cost the King anything, yet the yield of the calendar, its only source of income, was insufficient, and Leibniz spent much energy devising additional means of income.[122] None of his plans, however, except the silkworm monopoly, was accepted. There was personal friction between Leibniz and the two Jablonski brothers, one of whom was acting president and the other secretary. The building of the society's quarters went on slowly, so that the group had difficulty in finding a place to meet. The observatory was gotten ready somewhat more quickly; but there were no laboratories, and all experimenting had to be done at the homes of the Fellows.[123] Yet something was accomplished; the Fellows experimented, investigated, reported on scientific work, gathered information from correspondence and scientific journals and made magnetic observations in Russia; so that the report to the King in 1702 sounded rather promising.[124]

Diplomatic complications made Leibniz less popular in

[120] The second volume of *Miscellanea* was not published until 1723.

[121] U. H. Gundling, *Historie der Gelahrtheit, oder ausführliche Discourse*, Part III, p. 3206, says: "Almost everybody can gain appointment and the society rarely meets."

[122] Tax on traveling outside of Germany, monopoly of fire engines, missions, censorship over books, tax on paper, instalment of a lottery, etc.

[123] Harnack, *op. cit.*, I, 121.

[124] It spoke of astronomical observations, of finding a new method for making phosphorus, of a plan for publishing the *Miscellanea*, of getting funds for regulating weights and measures, and putting them on the decimal basis.

Berlin, especially after the death of the Queen (1705); yet he remained the moving spirit of the Academy. After a decline in 1706, there was a revival of work from 1707 to 1710, much of the energy being devoted to the publication of the first volume of the *Miscellanea*,[125] which was to establish the Academy in the world of science. Its contents are the best comment on the activities of the society. It is characteristically in Latin—Berlin did not dare to break so fundamentally with precedent as to adopt the vernacular. The work was published in three sections: the mathematical-mechanical section was represented by thirty-seven papers, the physical by fourteen, the literary by seven. This shows clearly that in Germany the mathematical-mechanical interest was predominating. Among the sixty articles, twelve were from Leibniz' pen—demonstrating how great an element he was in the Berlin Academy. Moreover, his papers were in all three sections; as Fontenelle said: "Leibniz appeared here in all his learned rôles—historian, antiquarian, etymologist, physicist and mathematician." The quality of the work as a whole was of such high grade that it admitted of no doubt that, owing to Leibniz' activities, the new science had, by 1710, found a home in Berlin.

After 1710 the society's relations to Leibniz became strained to an extreme. He was treated abominably by Jablonski;[126] he was neglected, his salary fraudulently denied him, and as the society, but for Leibniz, had really never been alive, it inevitably declined. Then it must not be forgotten that the throne of Prussia was soon occupied by Frederick William I, who hated science as mere babble and effeminateness, regarded Leibniz as a man, not useful enough even to use as a patrolman (*Schildwach*),[127] and who had the meanness to charge the Academy rent for its use of the observatory,[128] so that nothing but decline

[125] *Miscellanea Berolinensia ad incrementum scientiarum*, Berlin and Halle, 1710; Harnack, *op. cit.*, I, 160 ff.

[126] Harnack, *op. cit.*, I, 169 ff. Leipniz never was in Berlin after 1711.

[127] *Ibid.*, p. 183.   [128] *Ibid.*, p. 189.

could be expected. The splendid revival of the Academy falls outside of the scope of our discussion.

The Berlin Academy, even if it had come up to the expectations of Leibniz, would not have satisfied his desire for learned societies. For his ambition in its entirety was to found a series of co-operating societies in different localities which should be scattered ultimately over the entire globe. As a beginning we find him in 1703 trying to found a society in Dresden.[129] Tschirnhausen had planned in years past a mathematical-physical society in Leipzig. A similar project was now proposed by Leibniz to the ruler of Saxony and accepted by him. The society of which Leibniz was to be president was to obtain its financial support by a tobacco monopoly. In addition to the scope of the Berlin Academy it was to include the publication of statistical tables of disease,[130] to be a "house of intelligence and issue bills of mortality," and was to have the oversight of education.[131] Historical works,[132] a dictionary,[133] etc., were to be edited. On the eve of the adoption of this plan the outbreak of war frustrated the scheme; but nevertheless Leibniz must be counted among the potential founders of the Dresden Academy which came to life so much later.

Within the last five years of Leibniz' life he was seriously engaged in projects for two more scientific societies. Russia had always attracted him because it was yet *tabula rasa*,[134] had not yet acquired the taste for studies, and could from the start be kept from reiterating the worst errors of the "system." Moreover, its hugeness and proximity to Asia fired his imagination.[135] In 1711 he met the Czar and outlined to him his program of es-

[129] Fischer, *op. cit.*, p. 235.
[130] Foucher de Careil, *op. cit.*, VII, 226.
[131] Leibniz had been appointed counselor in the education of the son of the ruler (*ibid.*, p. 234).
[132] *Ibid.*, p. 220.   [134] *Ibid.*, p. 468.
[133] *Ibid.*, p. 272.   [135] *Ibid.*, p. 395.

tablishing learning in Russia; the study of *realia* must be foremost; laboratories and observatories must be founded; scientific expeditions to Siberia and China should be made;[136] extensive magnetic observations be arranged;[137] *dictionaria technica* devised,[138] in which the terminology of arts and crafts should be explained in words and pictures. All this could be done if supervised by a learned society such as he had established in Berlin.[139] In Peter the Great, Leibniz felt he had found that patron of science he sought for. Peter in turn gave him a pension of £1,000—but no society was established.[140] Yet the founding of the Academy of St. Petersburg (1724) can clearly be traced to Leibniz' suggestions.

The years 1712-14 Leibniz spent in Vienna, very much respected by all prominent men, intimate with Prince Eugenio[141] and the Emperor. Here again he worked for the establishment of an imperial society. This was planned on the broadest lines;[142] it was to be the center of a series of institutes for the building of machines, observatories, physical laboratories, minerological and botanical collections. Here an encyclopedia of knowledge, along Baconian lines, was projected in order that gradually the knowledge of the individual scientist was to become the common property of all men. He obtained an imperial protocol[143] for this foundation but, through the activities of the Jesuits, the plan was defeated, and the Academy of Vienna was founded one hundred and thirty years later.

Soon after this Leibniz died, presumably heart broken at

[136] *Ibid.*, pp. 519-46.
[137] *Ibid.*, p. 563.   [138] *Ibid.*, pp. 584 ff.
[139] "Muscovite Plan," *ibid.*, pp. 472 ff.
[140] Harnack, *op. cit.*, I, 182.
[141] Foucher de Careil, "Letter to Prince Eugene," *op. cit.*, VII, 312-15.
[142] Fischer, *op. cit.*, pp. 136-39.
[143] Foucher de Careil, *op. cit.*, VII, 339; "Kaiserliches Decret zur Errichtung der Universität," *ibid.*, p. 374.

the failure of most of his plans. As we contemplate the phase of his activities studied here, his ideals stand out perfectly clearly. Scientific societies are the only means whereby useful knowledge can be developed and spread. It was his ambition to create such scientific societies throughout all important cities, as far as culture and civilization reached; they were to communicate with one another and thus create a federation of learned men who would through science guide the destinies and civilization of mankind. It was a great idea, and so much in advance of his time that it can hardly be said to be accomplished even today to the extent which he had conceived.

We may say, in conclusion, that, at the opening of the eighteenth century, no place in Germany existed where experimental science was fostered and cultivated as it was in the Royal Society and Académie des Sciences. Indeed, through the efforts of Leibniz the need and importance of founding such centers for scientific work had been loudly proclaimed, and one such center in Berlin had been founded, but most of his words and advice remained unheeded. It is perhaps in consequence of this that Germany remained on a lower level in experimental science than France and England for many decades, and that, except in chemistry, the names of the pioneers of science are not to be found among the Germans.

CHAPTER VII

THE SCIENTIFIC JOURNALS

In studying the activities of the scientific societies in the various countries during the seventeenth century, it has been seen that their functions were really twofold; they created laboratories and observatories and did a great deal to encourage original work; but they often, in addition, undertook to publish, periodically, news of the work done under their auspices, and often the work of other learned men, in order to make it known as quickly and as widely as possible. These two functions of investigation and propaganda are not necessarily inseparable. Indeed, the scientific journal can be conceived of as the organ of an association of men who laid the entire stress upon the latter feature, and consciously divorced the encouragement of original research from their aims. For this reason a consideration of certain scientific journals may be included in our discussion of learned societies.

It is not surprising that the same century that gave birth to experimental science witnessed the development of scientific journalism. For the experimenter must know how far and by what methods fellow-workers have solved the problems in which he is engaged; he cannot dispense with an organ which shall bring him this information. The age which created for him laboratories and the milieu for scientific work was predestined also to supply this need.

The only means of scientific intercommunication in the early seventeenth century was private correspondence. Hence the great significance of such men as Mersenne,[1] Peiresc,[2] Collins,[3]

[1] See above, chap. v.   [2] See above, chap. ii.

[3] J. Collins, *Commercium epistolicum* (published by the Royal Society). *Op. cit.*, published Newtoni, *Opera*, IV (1782), 443 ff.

and Wallis,[4] who kept up a voluminous correspondence, and the necessity that such scientists as Huygens and Boyle should be in personal communication with other scientists. The unreliability of this form of communication is self-evident. It depended too much on friendly or hostile feeling, and at times on geographical contiguity, whether or not important discoveries reached the world. The numerous quarrels regarding scientific discoveries, as for instance between Torricelli and Pascal, Newton and Leibniz, Hooke and Huygens, best prove the insufficiency of such informal intercommunciations. In order to secure priority while keeping discoveries secret, ciphers were used. The right road to a solution of all these difficulties was clearly indicated when Denis de Sallo published in 1665 the first volume of the *Journal des sçavans*.

It is typical of the close relation of learned societies and the scientific journal that we find as their editors the same type of *Kulturträger* as in the members of the learned societies, men thoroughly in sympathy with progress, eager to add their quotum to the impulse of change, ready to help in the task of dispelling superstition and intolerance.

Such a man was Denis de Sallo.[5] He was a member of the Parlement of Paris, and belonged to the *coterie* of learned men with whom Colbert surrounded himself, in order to avail himself of their knowledge. It had been Sallo's habit in his extensive reading to engage several copyists and have them transcribe the most remarkable passages he came across in his readings; these he arranged so that he could obtain quickly information on any subject desired. It occurred to him that he might do for the public what he had done for himself. He submitted to Colbert a scheme of publishing weekly[6] matters of general interest. This idea, which seems so natural to us today, was hailed as a happy

[4] Cantor, *op. cit.*, III, 9.
[5] Hatin, *Histoire politique et littéraire de la presse en France*, II, 152.
[6] "Because news ages so quickly."

discovery, and the privilege of editing such a periodical under the name of *Journal des sçavans* was easily obtained.

The first number appeared January 5, 1665, published by Sallo under the assumed name of Sieur d' Hedouville. It was small in size, perhaps to emphasize its popular aims, or, as Zedler[7] suggests: "flüchtigblütige Franzosen haben Ekel vor Folianten." Its purpose is well stated in the address to the reader. It proposed, first, to give a catalogue and short description of books; second, to give obituaries of famous men and summarize their works; third, and most significant for us, the prospectus proclaims that the *Journal* will publish experiments in physics and chemistry which serve to explain natural phenomena, new discoveries in arts and sciences, useful machines, curious inventions of mathematicians, observations of the heavens, meteorological phenomena, and new anatomical findings in animals. The fourth point of the program was the publication of the principal decisions of tribunals and universities; the fifth, of current events in the world of letters.[8]

This extensive program the *Journal des sçavans* carried out. Sallo published it but four months,[9] for, on account of hostile Jesuit criticism, his license was withdrawn; nevertheless within this short time even the periodical proved its *raison d'être*. The direction of it was put into the hands of Sallo's former collaborator, l'Abbé Gallois,[10] who had lived with him, and whom we met as a member of the Académie des Sciences. For nine years he supervised the periodical,[11] then l' Abbé la Roque[12] became its publisher, until in 1702 the same l'Abbé Bignon[13] who rejuvenated the Académie des Sciences put the *Journal* on a new basis, edited by a board instead of by one man.

[7] Zedler, *Grosses vollständeges Universallexicon aller Wissenschaften und Künste*, p. 8.

[8] "Au lecteur," *Journal des sçavans*, Vol. I.

[9] Hatin, *op. cit.*, II, 174.   [10] *Ibid.*, p. 176.

[11] In the years of 1673 and 1674 it was practically extinct (*ibid.*).

[12] *Ibid.*, p. 178.   [13] *Ibid.*, p. 189.

THE SCIENTIFIC JOURNALS

In the publications of the first years we find that possibly one-third of the articles are on historical researches. The rest are concerned with the work of the scientific societies.[14] Especially the work of the members of "la compagnie qui assemblent à la bibliothèque du Roi"[15] fills the pages of the *Journal*. Thus, in 1667 we find a detailed account of Perrault's dissections,[16] of Pecquet's medical discoveries,[17] in 1669 a description of various mechanisms examined by the Académie,[18] in 1670 an article on Roberval's balance,[19] in 1672 the account of numerous experiments in congelation; Huygens and Mariotte communicated regularly with the *Journal*. In 1675 we find five reports of Huygens' pendulum clock,[20] in 1676 eight reports, in 1679 (a most scientific volume) five reports of happenings at the Académie.

Among the authors of scientific articles for the year 1665–66 were Bausch and Sachs, of the Collegium Naturae Curiosorum, Hevelius, Kircher, and Divini, the Italian maker of lenses. We find also Hooke's *Micrographia* reviewed at length. In 1667, and throughout the volumes of the succeeding years, the question of the transfusion of blood was discussed, indeed in 1668 there was a symposium on that question. In 1672 there was an account of Newton's telescope[21] and Huygens' criticism thereon; in 1675 an article by Leibniz, three articles on Boyle's experiments;[22] in 1676[23] articles on Grew and Malpighi, reports on Lémery's textbooks of chemistry. The issue of May 5, 1681, was entirely devoted to the question of comets. We have reports of experiments, reviews of scientific books, discussions of astronomical problems, extracts taken from contemporaneous papers,

---

[14] In the two years 1665 and 1666 eighteen articles of the *Philosophical Transactions* were published; in 1667, five; in 1668, six, containing Boyle's experiments; in 1675, eight; in 1677, fourteen; in 1678, six, etc.

[15] *Journal des sçavans*, II (1668), 386.

[16] *Ibid.* (1667), pp. 209, 230.

[17] *Ibid.*, p. 104.

[18] *Ibid.* (1669), p. 513.

[19] *Ibid.*, III (1670), 23, 26.

[20] *Ibid.*, Vol. IV (1675), *passim*.

[21] *Ibid.*, III (1672), 94.

[22] *Ibid.*, IV (1675), 204–84.

[23] *Ibid.*, V (1676), 3.

such as the *Philosophical Transactions,* the *Giornale di litterati,* the *Miscellanea,* etc.

This will be sufficient to indicate how closely affiliated this first scientific journal was with the world which the learned societies represent. It came to be the organ of intercommunication, not only between the members of the various societies, but also between the societies and the lay reading public both in France and the rest of the Continent. Its continued existence shows a wider interest in scientific questions at this time than is usually suspected. Indeed, so much was it identified with this interest that if there were a list of those who subscribed or read the *Journal des sçavans* we should be in possession of a record of that indefinable and fluctuating something called "popular interest in science."

How unmistakably the French periodical met a real need of the time can be gathered from the promptness with which the enterprise was copied in England, Italy, Germany, and Holland. The *Philosophical Transactions,* published March 1, 1665, two months after the *Journal des sçavans,* have been discussed in connection with the Royal Society. All subsequent scientific periodicals developed as imitations of these two—the *Journal des sçavans* was used as the model for periodicals which were to appeal to a broader reading public, the *Philosophical Transactions* with their bulky volumes of monthly publication came to be the standard for publications of scientific societies.

The first to copy the French periodical were the Italians in their *Giornale de litterati di Roma* (1668–79), continued as *Il giornale de litterati per tutto l'anno*—Parma (1668–90), and Modena (1692–97). In the Preface the editor, Michael Angiolo Ricci, whom we met as corresponding member of the Cimento, stated that he wished to give the Italians the benefit of a publication in the vernacular such as the French had.[24] The type of

[24] "Giornale de i dotti o eruditi o vogliamo dire de i letterati pochi anni sono introdotto in Parigi . . . . e stato, recevuto con multo applauso. . . . ."

their work seems to have been similar to that of the *Journal des sçavans*—indeed, the French paper took notice of many of their articles;[25] but the Italian periodical was often discontinued because it appealed to an infinitely smaller number of readers than its French model.

The next scientific journal in order of time was the German medical paper, *Miscellanea naturae curiosorum,* which has been discussed above.[26] Then there were the *Acta medica et philosophica hafniensia* (1673–80) edited by Thomas Bartholinus. In France several papers were published. The first (1679), *Nouvelles descouvertes sur toutes les parties de la médecine,*[27] published monthly for five years, by Nicholas de Blegny, a charlatan. When this paper was suppressed, he fled to Holland and with Gautier published the *Mercure sçavant,* which suggested the idea of publishing a periodical to Pierre Bayle. La Roque, the editor of the *Journal des sçavans* from 1675 until 1686, published from 1683 to 1686 a new periodical, as Hatin relates—in order not to let perish one thousand beautiful observations which could not find space in the *Journal des sçavans,* either on account of their length, or because they were less to the taste of the general public—under the name *Journaux de médecine ou observations des plus fameux médecines . . . . tirées des journaux étrangers ou des mémoires particulières.* There are references also to the *Collectanea medico-physica* (Amsterdam, 1680) published in the Dutch language. Thus the idea of a journal especially devoted to medicine was widely accepted.

Next in order of time comes a publication resembling more the *Philosophic Transactions* than the *Journal des sçavans*: the scientific journal of the Germans, the *Acta eruditorum,* published monthly at Leipzig from the year 1682.[28] Leipzig was then—as it is now—the center of the German book trade; it was therefore appropriate that it should produce a journal, part of whose

[25] *Journal des sçavans* (1668), *passim.*
[26] See above, chap. vi.
[27] Hatin, *op. cit.,* II, 255 ff.   [28] Hence also called *Acta Leipsiensia.*

function was to keep the world informed of the new books which were being published. As this paper represents the most essential contribution Germany made to the cause of science during the seventeenth century, it deserves careful study. The enterprise was supported by the Duke of Saxony. Otto Mencke, professor of morals and practical philosophy, in conjunction with a learned body, variously referred to as the *Collegium Gellianum* or *Leipsicum*,[29] undertook the publication of these *Acta eruditorum*. Mencke was assisted by numerous regular correspondents, representatives of all branches of learning,[30] not only throughout Germany, but also in England and Holland.[31] The publication appeared in huge folios, which show to Zedler that the German was *schwerbluetig-melancholicus*.[32] The articles were written in Latin, and no attention whatsoever was given to current events.[33]

The volumes contained learned articles and announcements —not criticisms—of learned books. These *Acta eruditorum* are a curious transitional product, reflecting past scholasticism while they are filled at the same time with the new knowledge. In the address to the reader Mencke pointed out the purpose of the publication, namely, that whereas the *Philosophical Transactions* dealt with experiment only, the *Journal des sçavans* and *Giornale de litterati* were too popular and literary and the *Miscellanea naturae curiosorum* limited to medicine and science, the *Acta eruditorum* planned to combine the important elements of all these and to make their direct appeal to the *erudite* class.[34] But how far the *Acta* differed in spirit from the *Philosophical Transactions* is clearly shown by a perusal of the Index volume[35]

[29] Gundling, *op. cit.*, Vol. III, p. 3210, Sec. 4; Prutz, *op. cit.*, p. 275.
[30] Prutz, *op. cit.*, p. 282 (among these was Leibniz).
[31] Cantor, *op. cit.*, III, 201.
[32] Zedler, *op. cit.*, p. 8.
[33] Prutz, *op. cit.*, p. 279.
[34] "Lectori," *Acta eruditorum*, Vol. I.   [35] *Ibid.*, Indices generales.

where special headings are allowed, not only for medicine and mathematics, but also for law, for "theological subjects," and for references to the Bible (*Locorum scripturae*), the editors being orthodox in their religious convictions.[36]

A rapid perusal of the first volumes however shows that the *Acta* were in *Fühlung* with the new Science. The first volume opens with Nehemiah Grew's work; Boyle, Sydenham, Papin, Borelli, Leeuwenhoek, Leibniz, Sturm, Bernouilli, Hevelius, were represented in its pages,[37] and the following volumes continued to count the most progressive scientists among their contributors,[38] and in their pages are found many statements of interest in the history of science. Indeed the *Acta* stand in the

[36] The following table gives an idea of the relative stress laid on various subjects in the volumes (1683–1700):

	I	II	III	IV	V	VI	VII	VIII	IX	X	XI	XII	XIII	XIV	XV	XVI	XVII
Theology	35	50	56	71	55	65	56	48	54	59	56	40	38	46	50	48	43
Law	16	19	11	24	9	7	16	10	13	5	5	12	15	13	6	14	12
Medicine and physics	43	48	47	48	48	32	33	26	19	19	18	28	20	24	18	21	26
Mathematics and astronomy	23	33	31	28	30	16	25	24	15	29	22	26	33	36	22	24	17
History and geography	24	31	28	72	30	34	40	27	16	30	35	34	40	45	26	29	48
Philology	27	41	44		32	30	17	24	30	37	41	46	36	32	41	30	45

[37] Vol. I was translated into French. Published in the Haag, 1885.

[38] I adjoin a table with the names of the most prominent contributors and most interesting articles of the first sixteen volumes:

Vol. II. Borelli, Leeuwenhoek, Bernouilli, Descartes, Hevelius, Sturm.

Vol. III. Halley, Boyle, Grew, Pascal, Sydenham, Hevelius, Huygens, Sturm.

Vol. IV. Leeuwenhoek, *Anatomy and Contemplation of Several Invisible Secrets of Nature*. Waller's translation of the *"Saggi" of the Cimento*. Articles of Bertolimus, Bernouilli, Boyle (description of apparatus), Grew, Leeuwenhoek. Announcement of Swammerdam's book and of Sturm's publication of *Academia Altorfinna*.

Vol. V. Leeuwenhoek, Papin, Redus, Bernouilli, Campani, Cassini, Flamsteed, Wallis.

Vol. VI. Report of the method of studying magnetic declination, found by the French Academy; review of Malpighi's work; Ray and Willoughby, *History of Plants*; Boyle, Halley, Papin.

Vol. VII. Review of Newton's *Principia* in eleven folio pages (Newton is

very center of the scientific life of Europe—"Gelehrte schrieben für Gelehrte."[39] If we ask what they accomplished, we find that by proposing problems and recording solutions they constituted a forum for the exchange of ideas. They published works of scientists, as for example, Tschirnhausen's. They reviewed the scientific work of foreign nations, which was their main purpose,[40] and thus made Germany acquainted with the scientific accomplishments of the world. During a decade when Germany, through that lack of learned societies which Leibniz lamented, apparently lagged behind, they tried to collect what there was of German scholarship and inform the outside world of German scientific thought.[41] They seem to have excited admiration abroad;

---

        called "eximius nostri temporis Mathematicus"); articles of Boyle, Papin, Bernouilli, Cassini, Weigel.

Vol. VIII. Leeuwenhoek, Bernouilli, Papin, Sturm.

Vol. IX. Huygens' theory of light, Bernouilli, Papin and Reisellius (on a model of a steam engine), Sturm.

Vol. X. Bernouilli (on effervescence and fermentation), Boyle, Locke, Huygens, Hevelius, Bernouilli.

Vol. XI. A review of the mathematical and physical publications of the Académie des Sciences, published in Haag, 1692; Cassini, *Moons of Jupiter;* Bernouilli, *Caustics and Cycloids.*

Vol. XII. Boyle, Leibniz, Ray, Bernouilli, Campani, Cassini, Halley, Huygens, Leibniz, Weigel.

Vol. XIII. Continuation of the report of the Académie des Sciences; Bernouilli, *Mathematical Meditation on Motions of Muscles;* Ray, Jacob and John Bernouilli, Cassini, Viviani, Leibnitz (on the applications of the new differential calculus).

Vol. XIV. Leibnitz, Papin, Sydenham, Tschirnhausen, Tournefort, Bernouilli, Cassini, De La Hire, Hôpital, Huygens, Mariotte, Roberval.

Vol. XV. Grew, Tschirnhausen, Cassini, Leibniz, Wallis.

Vol. XVI. Malpighi (posthumous), Sturm, Leibniz, Newton (solution of problem proposed by Bernouilli), Tschirnhausen.

[39] Prutz, *op. cit.,* p. 286.

[40] *Acta eruditorum,* Vol. III, Preface.

[41] Tschirnhausen said his main purpose in publishing in the *Acta* was to come to the notice of the Académie des Sciences and thus to obtain a pension from Louis XIV (Cantor, *op. cit.,* III, 113).

"I found them so judicious, so exact, so diverse"—says Pierre Bayle—"that they surpass their great reputation."[42] To the historian they furnish a valuable record of the great scientific achievements of the seventeenth century; and their pages bear witness to the high work done in the scientific societies.

The only periodical which in any sense rivaled in popularity and influence the *Journal des sçavans* was the *Nouvelles de la république des lettres*. The famous Pierre Bayle, residing in Holland as professor of philosophy and history at the University of Rotterdam, decided in 1684[43] to publish this new periodical. While its name suggests a purely literary publication, it is in the type of its articles a popular scientific journal. In the Preface, Bayle appropriately said that while the idea of the *Journal des sçavans* had been copied from nation to nation, from science to science, by physicians and chemists, Holland alone had not done its share, although on account of the liberty of the press, works from every quarter were published there. He asked his readers to communicate new discoveries to him; and announced that the periodical would publish *Eloges* of scientists. He characteristically added that this would be done regardless of the religious persuasion of the scientists, with a view only to their scientific reputation; for in the republic of letters all learned men should regard one another as brothers. Bayle further promised that the *Nouvelles* would inform the public of the happenings at the various academies.[44] Referring only to the scientific element in the volumes, we find in their pages Papin's experiments on the circulation of blood[45] made in Venice, Wallis' and Boyle's[46] books; reviews of Leeuwenhoek's[47] article on the generation of

[42] *Nouvelles de la républiques des lettres*, I, 1.
[43] Hatin, *op. cit.*, II, 222.
[44] *Nouvelles de la république des lettres*, Vol. I, Preface.
[45] *Ibid.* (May, 1686), p. 563.
[46] *Ibid.* (June, 1684), pp. 76 ff.   [47] *Ibid.* (September, 1684), p. 118.

man written to Christian Wren—published in Latin so as not to shock the lay reader—an article of the *Journal des sçavans* about siphons[48] "invented by a man without knowledge of Latin or aid of any master"; articles on anatomy from the *Melanges curieux*,[49] etc. The paper was published for only three years but nevertheless greatly advanced the cause of periodical scientific literature. It was continued by H. Basnage de Beauval in the *Histoire des ouvrages des savants* (1687–1704) and by Leclerc in the *Bibliothèque universelle et historique* (1683–93),[50] where there are found publications of Leeuwenhoek and Hooke and Willoughby. On the whole, the scientific element in these later volumes is less apparent.

In the last decade of the seventeenth century, periodicals copying the *Journal des sçavans* and *Nouvelles de la république* multiplied; two of these were *Dépêche du Parnasse ou la gazette des savants* (Geneva, 1693),[51] and *Nouveau journal des savants dressé à Rotterdam*,[52] published by Etienne Chauvin, a French fugitive (Berlin, 1696–98).

It will be evident from this short survey that the cause of scientific periodicals was victorious by the end of the seventeenth century. They were of the greatest assistance in advancing the idea for which the learned societies stood. Publications of the type of the *Journal des sçavans* and *Nouvelles de la république des lettres* reached a much wider circle of men and brought the message of scientific thought and inquiry to many whom the learned societies or the more learned publications could never have reached. Periodicals of the type of the *Philosophic Transactions* and the *Acta* were invaluable; for thus there were created channels through which the *savants* and *naturae curiosi*

[48] *Ibid.* (February, 1865), p. 230.
[49] *Ibid.* (July, 1685), p. 336.
[50] Hatin, *op. cit.*, II, 251.
[51] *Ibid.*, p. 256.   [52] *Ibid.*, p. 257.

could publish their ideas, communicate with one another, and be informed of the progress of science, and in this way those objects for which the learned society had been founded were promoted. Periodicals of the type of *Miscellanea naturae curiosorum* are of special interest, as they foreshadowed the publications of more specialized societies of a later time. The scientific journal must be thought of, therefore, as an invaluable instrument which the seventeenth century created partly in the service of, partly cooperating with, the learned societies.

PART III

THE LEARNED SOCIETIES AND THE UNIVERSITIES

CHAPTER VIII

SCIENCE IN THE UNIVERSITIES

We shall now turn to a cursory study of the attitude taken by the universities in the seventeenth century toward the experimental sciences. The European universities in 1600, however they might differ from one another in detail, had many common characteristic features which were preserved unchanged from the medieval scholastic period. The beginning of all university study was the four years' course leading to the Bachelor's degree, comprising grammar, logic, rhetoric, and the four mathematical subjects—arithmetic, music, geometry, and astronomy.[1] Astronomy was, of course, not the science as we conceive it, with observatory and telescope, but was based on a study of Ptolemy's standard treatise. Three additional years spent in the study of *utraque philosophia*, of natural philosophy, moral philosophy, and metaphysics, brought the Master's degree. Natural philosophy or physics meant the reading of Aristotle's books. These requirements for the degrees differed in various universities, but the amount covered was always about the same—for the most part a study of the works of Aristotle. After obtaining the Master's degree, the student might attach himself to the faculty of theology, law, or medicine to attain, in a time varying from four to seven years, the professional degree. The theological faculty was of the highest dignity, suggesting the ecclesiastical auspices under which the universities had often developed. Most administrative offices of the university were held by the theologians; and the required adherence to definite religious views put the church's stamp upon the whole institution. The faculties of philosophy and of law were closely allied

[1] *Register of the University of Oxford (1571-1622)* (ed. Andrew Clark), II, Part I, 225.

to that of theology; that of medicine stood apart, and was therefore the least significant. All instruction and even communication between the students was in Latin. On account of this, and a series of laws specially protecting those attending the universities, a hedge seemed set about them, unsurmountable to the *vulgus profanum*. The instruction by the professors consisted of "lectures" in the literal sense of the word, a reading of the prescribed text with such comments and explanations as they chose to add. The textbooks were mainly Aristotle's, or commentaries upon him. Mathematics consisted in the explanation of Euclid; medical instruction in a commentary on Galen or Hippocrates.[2] The professors' function being thus limited, there was little attempt at specialization, and often widely different subjects were "read" by the same instructor. The student, besides attending these lectures, had to defend theses, selected by himself or by the professor, at numerous disputations, private and public. These disputations were never supposed to bring in new subject matter; they were only to give the student a chance of combining and recombining, of practicing and exercising, what he had acquired from Aristotle or other authorities. This feeling of the finality and sufficiency of what had been discovered by the standard Greek writers was at once the most characteristic and most pernicious feature of the system. At these disputations the professor presided, and saw that the arguments proceeded along the strict lines laid down by Aristotle's logic.[3]

[2] Dr. Theodor Puschmann, *A History of Medical Education from the Most Remote to the Most Recent Times*, pp. 333 f.

[3] Some rather startling subjects of disputations at Oxford were:
1601. "An recte fecerint Graeci et Persae qui inter pocula deliberaverint?"
1602. "An polita studiosorum scientia sit plebeis communicanda?" (Neg.)
1605. "Utrum praestet in omnibus scientibus mediocrem esse quam in una aliqua singularem?" (Neg.)
1608. "An quisquam sibi stultus videatur?"
1608. "An foemina sit idonea auditrix moralis philosophiae?" (Neg.)
1614. "An animo sit tabula rasa?" (Affirm.)
1621. "An cometae sunt mutationum in republica praesagi?" (Affirm.)

SCIENCE IN THE UNIVERSITIES

Throughout the universities there reigned a spirit opposed to freedom of thought which has received its masterful expression in the *Ratio studiorum* of Acquaviva, the Jesuit general:

The teacher is not to permit any novel opinions or discussions to be mooted; nor to cite or allow others to cite the opinions of an author not of known repute; nor to teach or suffer to be taught anything contrary to prevalent opinions of acknowledged doctors current in the schools. Obsolete and false opinions are not to be mentioned at all even for refutation nor are objections to received teaching to be dwelt on at any length. . . . . In philosophy Aristotle is always to be followed and St. Thomas Aquinas generally. . . . .[4]

Such traces of scientific interest as existed were to be found in the faculty of medicine. Its professor taught botany so far as it related to medicine; some chemistry in connection with the use of drugs. Anatomical and physiological instruction consisted of the verbal analysis of Galen—but a barber might be present who would roughly dissect an animal or, on extremely rare occasions, once or twice a year, the body of an executed criminal which had been handed over to the medical school. These dissections would serve as concrete illustrations of what the professor was teaching, not as tests by which the truth of what he was stating might be proved.[5]

Everything connected with the external life of the university —the appearance of professors and students, the system of examinations, the formalities of obtaining the degree—were matters of most punctilious regulations.

Such institutions could only become the home of modern scientific study by adopting fundamental reforms, which may be classed under the following heads:

1. The universities must needs become secularized; for only in a non-ecclesiastic atmosphere could the sciences thrive.

2. The entire mode of instruction must needs be reformed. It was essential to substitute original observation for the old habit of reading standard texts and conducting disputations.

[4] Quoted in *Encycl. Brit.* (11th ed.), "Jesuits."
[5] Puschmann, *op. cit.*, pp. 321 ff.

3. The studies under the faculty of philosophy must be reformed by minimizing the emphasis on the study of logic, metaphysics, and ancient languages.

4. The study of the various sciences must be established as independent disciplines. This would imply necessarily the establishment of a body of more specialized professors.

5. Experimentation and observation as a regular method of study must be adopted. Hence the necessity of establishing places for such study, namely, laboratories, botanical gardens, observatories, mineral and zoölogical collections, etc.

6. Current scientific discoveries must be incorporated into the textbooks used by the student.

7. The study of medicine must be reformed, and established entirely on the basis of anatomical experimentation and study at the sick bed.

8. The faculty of theology must be relegated to a secondary place. The faculty of philosophy should be supplemented by, or metamorphosed into, a faculty of science.

9. The vernacular must be substituted for Latin as the vehicle of study and teaching.

10. University interest must reach out to the objects of everyday life.

11. There must be freedom of conscience and of thought in matters of philosophy and scientific inquiry.

12. There must be freedom of the press.

Did any of these conditions prevail at any universities in the seventeenth century, and, if not, were such changes even suggested? This will be the subject of inquiry in this chapter.

As the university development varied considerably in the different European countries, I shall take up separately certain matters of interest to us regarding the university instruction in Italy, France, Germany, England, and the Netherlands. But I must confess that in most cases I have been disappointed in the amount of material I could discover. I have no reason to think, however, that, were additional information available, it would essentially alter the impressions one gets of the rôle of the universities during this period.

ITALY

The Italian universities in 1600 present many features which would point to progress in the right direction. They were either under the control of wealthy municipalities as, for example, the University of Padua under Venice, that of Pisa under Florence; or they were like the University of Bologna, so rich and independent that they were relatively free from church control. Affiliated with the liberal spirit of the North Italian city republics, they had adopted a most liberal policy toward professors and students. Neither creed nor country was a bar to advancement; and foreigners of every nation were to be found among their professors. Padua was most frequented by Germans, especially for medicine, owing to the close commercial relations between Venice and Germany, and Germans are found among the rectors there. Bologna also, with its famous law faculty, and Siena had their large share of German students.[6] English students were attracted in considerable numbers, and biographies of prominent men of this time indicate that many spent their *Wander-* and *Lehrjahre* in the great institutions of Northern Italy. This external sign of their flourishing condition is substantiated by a consideration of their internal policies. The wealth of the universities permitted a high degree of specialization, even the existence of several chairs in a single field. A "call" to Padua or Pisa was deemed the highest honor in the professional world.

In the introduction of the experimental study of medicine Italian universities were pioneers.[7] Here the baleful separation

[6] Franz Eulenburg, *Die Frequenz der deutschen Universitäten* . . . . *Abhandlungen der Königl. Sächs. Gesellschaft der Wissenschaften* ("Philologisch-historische Klasse," Vol. XXIV¹), LIII², 124.

[7] Foster, *op. cit.*, p. 104.

between surgery and medicine had never been as complete as in other countries.[8] Here, as we saw, Vesalius first insisted on the dignity of anatomical investigation. Padua had the first botanical garden and the first anatomical theater of the world,[9] and here as early as 1578 clinical instruction at the bedside and the postmortem examination, "if the season of the year permitted,"[10] was begun. But these most progressive features of medical instruction declined before 1600. Yet it is interesting to note that even at the most liberal University of Padua, Vesalius, for a long time did not dare to teach what he really saw, but accommodated his statements to the accepted dogmas of Galen. When he finally allowed himself to state the facts, and even offered to prove his statements publicly in the dissecting theater, the opposition was so persistent that in a fit of passion he tore up his manuscripts and left the university for the court of Spain.[11] Similarly Galileo, lecturing on the theory of the planets, persisted in expounding it according to the Ptolemaic, not the Copernican system, although the latter was not as yet condemned. He wrote (1597) to Kepler that while he believed in the Copernican hypotheses, his function as a professor demanded of him nothing but to hand down the accepted opinions of the past.[12] This, in a word, is a condemnation of the entire system.

The condemnation of Galileo certainly served as a severe check to the modern study of astronomy in the universities. During the seventeenth century, however, experimental physics was taught in Italian universities, even at the Jesuit school at Bologna. The Jesuits Riccioli and Grimaldi were skilful experimenters,[13] and—a veritable triumph for Galileo's methods—

[8] Puschmann, *op. cit.*, p. 336.
[9] Minerva, *Handbuch der gelehrten Welt. Die Universitäten und Hochschulen, ihre Geschichte und Organization*, I, 326.
[10] Puschmann, *op. cit.*, p. 332.   [11] Foster, *op. cit.*, pp. 14 f.
[12] Quoted in Strauss, *op. cit.*, Vol. XV.
[13] Rosenberger, *op. cit.*, II, 130.

SCIENCE IN THE UNIVERSITIES

tried to refute him by experiment. Most of the Cimento scientists were, before or after their short-lived membership in the Academy, members of the faculties of Pisa, Padua, or Messena.[14] A chair of theoretical medicine was founded for Malpighi in Bologna. Sanctorius (1561–1626), the first to use scales and thermometer for medical purposes, taught at Padua; Aselli, the discoverer of the lymphatic system, taught at Pavia.[15] Such laboratory work as was done may have been performed mostly at the houses of professors. Foster reports that B. Massari, professor of medicine at Bologna in 1650, gathered his nine students into a club, Chorus Anatomicus, and carried on dissections, experimenting even on living animals.[16]

The only instance of the establishment of a regular university laboratory has a peculiar importance for the student of learned societies. Count Ludovico Ferdinando Marsiglio,[17] a science-loving amateur in Bologna, had collected in his travels many books, mechanisms, and instruments, in which he invested most of his inheritance, and at his house convened an academy of experimenters, Philosophi Inquieti. In 1690 he bequeathed his house and laboratories to the University of Bologna, to be perpetuated as an institution of research (Instituto della Scienze) for public instruction, to teach practical science where hitherto only theoretical science had been taught. A library, laboratory, physical cabinet, and observatory were established, employing an astronomer, a mathematician, an experimental physicist, a chemist, and a librarian. While the formal organization of this institute was not completed until 1711, it belongs by its donation (1690) to the seventeenth century and is one of the rare instances of the adoption of the experimental method by a university.

[14] See chap. iii.
[15] Foster, p. 48.
[16] *Ibid.*, p. 87.   [17] Mazzetti, *op. cit.*, pp. 64 ff.

FRANCE

In none of the larger countries do the universities in the seventeenth century offer as unpromising a spectacle as in France. The voluminous book of Jourdain, *Histoire de l'Université de Paris au 17ᵉ et au 18ᵉ siècles*, gives a fairly complete picture of the University of Paris. And in speaking of that university, we are practically discussing those of all France, since the provincial universities, with one exception, were very insignificant.[18]

The University of Paris was, at the end of the sixteenth century, by far the most important in Europe, counting thirty thousand students, more than all the Italian universities together. It obtained new statutes in 1600 from Henry IV, and these give an excellent idea of the status of the university at the opening of the century.[19] Only Catholics were admitted; the first function of the rector of the university was to impress the importance of religious duties upon the students. The costumes were prescribed; all conversation even among the students was to be in Latin. For the students of philosophy a two years' course of study was laid down, mostly given over to Aristotle: in the first year his *Logic*, in the second his *Physics* and *Metaphysics;* in the morning Euclid was to be read. In reading Aristotle, more attention was to be paid to the thoughts than to the text; "a strange conflict," says Jourdain, "between the scholastic veneration of Aristotle and the modern critical spirit." A definite number of disputations were required. The church ceremonies connected with taking the degree were regulated in every detail.

The medical course consisted of two years' study, though a

[18] The exception was the University of Montpelier, next to Padua the most famous medical school of Europe. I have not been able, however, to obtain any account of the development of this important institution during the seventeenth century. The fact that it had an anatomical theater in 1598 suggests that it was advancing at the same pace as other great medical schools (Puschmann, *op. cit.*, p. 329).

[19] Charles M. Jourdain, *Histoire de l'Université de Paris au 17ᵉ et au 18ᵉ siècles*, pp. 12 ff.

SCIENCE IN THE UNIVERSITIES

four years' "residence" was required. The student must study the "virtues" of plants; read five books of Galen on their properties, and make excursions to the botanical garden. He must be present at two dissections. The professors, dressed in long robes with cap and *chausses d'écarlate*, were to give two courses: one in the morning on physiology; one in the evening on pathology and therapeutics, reading Hippocrates and Galen. Two doctors were to visit the pharmacies and read to the apothecaries a course on *materia medica* and pharmacology; two were to read to surgeons and barbers. Then there followed a most detailed account of examinations. To these, surgeons, of course, were not admitted,[20] and an additional statute of 1607 emphasized this point especially.[21]

The condition of the university in the hundred years following the appearance of these statutes must be regarded as one of utter stagnation. The most important works edited under the auspices of its faculties were a new edition in Greek and Latin of Aristotle, and an edition of Hippocrates and Galen.[22] The use of the vernacular was suppressed whenever possible. Camus was prohibited from lecturing in French in 1624,[23] but the theses of a certain Alexis Trousset, submitted to the faculty in French, were sustained by request of the Queen Mother.[24] The censorship of books—a function of the university—was most strict. The death penalty was prescribed if the book were printed without authorization (1626). Of course, there was a systematic suppression of the liberty of thought. Sieur Jean Bitaud offered the thesis that "Aristotle's teaching in regard to the four elements

[20] *Ibid.*, p. 15.

[21] "The science is not for those who have only hands; this must be left to the judgment of physicians" (*qu'ils doivent laisser à juger aux médecins*) (Puschmann, *op. cit.*, p. 337).

[22] Jourdain, *op. cit.*, p. 136.

[23] G. Compayre, *Histoire critique des doctrines de l' éducation en France depuis le sezième siècle*, p. 422.

[24] Jourdain, *op. cit.*, pp. 106 f.

was wrong"; the thesis was torn up, the author forced to leave within twenty-four hours, and his license to teach canceled.[24a] But still worse—a decree was passed forbidding anyone at the peril of his life to hold or teach anything against the ancient authors, or to hold disputations against what was accepted by the faculty of theology.

The most famous and characteristic instance of the university's dealing with "New Thought" was its treatment of Descartes' teachings.[25] Descartes had dedicated his *Méthode* to the Sorbonne,[26] predicting that, if their approval could be bestowed on his writings, the arguments whereby he had sought to demonstrate the truth of the existence of a God and the immortality of the soul would then find such acceptance by both the learned and the scientific world that atheism would disappear from among civilized mankind.[27] Rome had not condemned him, the government had not proceeded against him, but the school of theology of the University of Paris found that Descartes' ideas were in conflict with orthodox teaching, i.e., that his fundamental notions of the supremacy of intellect were at variance with the doctrine of transubstantiation. Hence through their activity Descartes' works were put on the Index (1663) *donec corrigantur*. The funeral oration to be held by a former rector of the university upon the transference of Descartes' body to Paris was interdicted (1667). In 1669 the chair of philosophy was filled by a candidate who upheld the thesis of the "excellence of peripatetic philosophy against the novel teachings of Descartes." In 1671 the Archbishop of Paris called to his presence all members of the four faculties and addressed them as follows:

It has come to the ears of the King that views, censured by the faculty of theology and forbidden by the Parlement, are spreading not only within

---

[24a] Jourdain, *op. cit.*, pp. 106 f.
[25] *Ibid.*, pp. 233 ff.
[26] "Cum tanta inhaereat omnium mentibus de vestra Facultate opinio, tantaeque sit auctoritatis Sorbonae nomen, ..."
[27] "Ut Athei contradicendi animum deponant."

the university but everywhere. He desires to stop the advance of opinions which might bring confusion into the explanation of the mysteries of the church. It is the duty of the professors to look to it that no doctrine be taught nor allowed to get into the theses except that admitted in the regulations and statutes of the university.

This was accepted by the faculties without a dissenting voice.[28] The opposition to Descartes was definitely formulated in 1678. Here we find, "in physics it is forbidden to deviate from the principles of the physics of Aristotle [and this almost one hundred years after Galileo's experiments] accepted in the colleges, and attach oneself to the new doctrines of Descartes." Then follows the enumeration of nine objectionable points in Descartes' doctrine, among them the error that actual extension is the essence of matter, as this would interfere with the theory of the Eucharist, etc.

In 1685 Gassendi's teachings were also forbidden,[29] and in 1691 another list of errors of Cartesianism and Jansenism was drawn up, including among the errors Descartes' doctrine that the Christian may doubt everything, even the existence of God, and may deduce God's existence from reason, and require that his faith agree with reason.

In enumerating these instances of opposition to free thought, it seems strange that the university did not formally condemn the "Copernican hypothesis." It came, indeed, very near to this in 1631 through the influence of the astronomer Jean Baptiste Morin (1583–1656); and it was due to the defense of the system and refutation of Morin's reasons by facts advanced by Gassendi that the Sorbonne refrained from re-enforcing the decree of the Roman Inquisition.[30]

The university was most intolerant in religious matters.[31] When in 1638 the rector was informed that the medical doctorate was about to be attained by a Protestant, son of the phy-

---

[28] Jourdain, *op. cit.*, p. 235.
[29] Jourdain, *ibid.*, p. 269.
[30] Poggendorff, *op. cit.*, p. 303.
[31] Jourdain, *op. cit.*, p. 135.

sician of the Duke of Orléans, he formally objected and insisted on the exclusion of all heretics from candidature.

The university was equally intolerant of all attempt to reform the methods of education.[32] There had sprung up a number of centers of learning which claimed to instruct the youth in Latin, Greek, and the sciences in a less laborious, perhaps also less thorough, way than the university. Against these the university waged a protracted war, in which King and Parlement were on its side. The reasoning of the university against these "innovating colleges" is an odd bit of sophistry:

> The chemists are working in vain, for they will never give us gold resembling that created by nature. Fruits and flowers whose growth is forced, will never have the taste and odor of those matured at their natural season. Similarly in studies, nature's course must be followed. The time she allots must be well applied, but results must not be forced violently. In view of this the course of instruction has been instituted.[33]

It never occurred to the speaker that his analogy between natural forces and the traditions of instruction was somewhat weak.

The vitality of the university was exhausted throughout the first half of the century in battling against Jansenism and against Jesuit control, to which it gradually had to submit; thus the history of this, once greatest of universities, is one of insignificance and decline. As its historian Jourdain admits, "it records no considerable events or important improvements or great activity."[34]

The new spirit and the new science of the seventeenth century left hardly any traces upon the university curriculum. Chemistry was interdicted, physics remained Aristotelian; bot-

[32] *Ibid.*, p. 238.   [33] *Ibid.*, p. 240.

[34] *Ibid.*, p. 139. The criticisms of Louis XIV and Colbert seem justified. Colbert insisted on the lay charter of the university, deducing this from legal reasons; Louis XIV, with instinctive dislike of ecclesiastics, and tendency to self-aggrandizement, insisted that education was a civil function, belonging to the state; that the university should look to the King, and not to the Pope, as its head. He had in mind (1667) a series of reforms along these lines, including all universities. But these plans never reached execution (*ibid.*, pp. 228 f.).

any only was somewhat cultivated.[85] A botanical garden was established, and in 1646 a course in botany was given. The greatest reform was the introduction of polyclinical instruction in the medical faculty in 1644, which was due to what might almost be called an accident.[86]

Slight as was the scientific progress in the university, there existed in Paris an institution which seemed predestined as a home of the new sciences. In 1518 Francis I had founded the Collège de France on the model of the University of Louvain, to establish a home for humanism, then bitterly opposed at the University of Paris. Instruction was to be free of charge and open to the public; no degrees were given. Originally there were two professors of Greek and two of Hebrew, later there were added chairs of medicine, mathematics, philosophy, eloquence, surgery, botany, Arabic and canon law.[37] Here Peter Ramus had

[85] *Ibid.*, p. 21 n.

[86] Puschmann, *op. cit.*, pp. 412 f.: "This measure was adopted at the instance of Theophraste Renaudot. This clever and enterprising man, who founded the first loan office and the first *bureau d'addresse* in Paris and who also edited the first French newspaper, the *Gazette de France,* organized in conjunction with some medical colleagues an institution of the nature of a dispensary for relieving poor patients gratis. This brought him the ill will of the medical faculty, with whom he lived in a continual feud, as he would not fall in with the exclusive party spirit which animated that body. To such a point did they carry their opposition that, after the death of Renaudot's patron, Richelieu, his polyclinic, which had been a source of so much benefit to the poorer population, was closed. But now the medical faculty itself had to assume the duty of maintaining a similar institution. . . . .

"It was arranged that six doctors, three old and three young, should be commissioned to examine and supply with medicines twice every week in the École de Medecine such patients as applied; their services were to be gratis. Surgical operations were either to be undertaken by these doctors or else by a skilful surgeon. . . . . Poor patients not in a condition to come to the clinic were visited gratis in their dwellings. The 'bachelors,' in other words the senior students of medicine, were obliged to attend the polyclinical consultations, and were employed there to write down the prescriptions dictated by the doctors and to render other services. . . . ."

[37] This college was founded with only the promise of a building, and the professors taught at various places (Jourdain, *op. cit.,* p. 112).

taught mathematics and his traditions were cherished; Gassendi had held the professorship of mathematics until 1647,[38] and Roberval had occupied the same position on the interesting condition that he should resign to anyone who could furnish better solutions to the problems proposed—but he held it throughout his life. Somehow, this institution declined in the seventeenth century; the teaching posts became well-nigh hereditary,[39] and it seems to have been of little consequence in the later decades.

After this survey we can but repeat that in France the prospects of science in the university and even in the Collège de France were hardly brighter in 1700 than in 1600.

GERMANY

For the study of German universities in the seventeenth century there exists a masterly treatment by Franz Eulenburg.[40] On the basis of a close examination of the matriculation lists he reaches the most valuable generalizations for this period.[41]

The number of German universities rose from seventeen to thirty-nine in this period, but this remarkable increase was not the consequence of increased enthusiasm for learning. It was due in part to the establishment of centers of Lutheran and Calvinistic teaching, in part to the decentralized condition of Germany, to the fact that the ruler of even a small territory wanted

---

[38] R. Wolf, *op. cit.*, p. 327.   [39] Jourdain, *op. cit.*, p. 115.

[40] Eulenburg, *op. cit.*, No. 11. The only disadvantage for us would seem that in periodizing the history of German universities he puts the second period from 1550 to 1700, thus embracing the fifty years before the seventeenth century; but this is less important in view of his explanation that the changes which affected the universities, as they were to continue for a century and a half following, viz., the establishment of the Protestant universities, the commencement of Jesuit control of the Catholic universities, and the introduction of humanistic teaching, had occurred about 1550.

[41] Eulenburg, *op. cit.*, p. 191 n., complains of the difficult task of studying the German universities: "Leider fehlen auch hier bisher die geeigneten Vorarbeiten; Paulsen behandelt nur den gelehrten Unterricht, Tholuck nur die Theologie, Stöltzel nur die Rechtswissenschaft."

his own university.[42] Hence many of these German "universities" were exceedingly poor,[43] had the smallest possible number of professors and consequently an entire lack of differentiation of instruction. The continued frequent migrations of universities, such as occurred in earlier centuries, indicate the absence of apparatus to be moved.

Of the thirty-nine universities, twenty-three were Protestant, sixteen Catholic; all of the latter were in the hands of the Jesuits, with the exception of Salzburg which was under the control of the Benedictines.

The religious element was dominant both in Protestant and Catholic institutions, more if anything, in the former; all administrative offices were in the hands of the theologians, the conferring of degrees was a religious exercise. The theological faculty in the Protestant university was often so prominent a feature that the institution was looked upon as a "theological seminary." Emphasis was always placed on religious orthodoxy. In the statutes of the University of Tübingen, it was provided (1601) that "no one should be admitted to the degree concerning whose true religion there may be any doubt."[44] Julius of Brunswick, founder of the University of Helmstadt, declared: "He who is not in harmony with the teachings of the church is not to be tolerated; it were better, he go to the devil, than that he defile church and school."[45] In most universities, professors and students were re-

[42] The founding of the following universities is due to political ambitions: Jena, Helmstadt, Giessen, Kassel, Densburg, Kiel; to confessional reasons: Dillingen, Würzburg, Paderborn, Osnabrück, Altdorf (*ibid.*, p. 80).

[43] The most prominent of the universities were, in 1600, Wittenberg, Leipzig, and Helmstadt; after the Thirty Years' War, Leipzig, Jena, Wittenberg (*ibid.*, p. 81).

[44] "Nullo tempore de cujus sincera religione dubitetur ad professionem eligitur" (August Tholuck, *Das akademische Leben im siebenzehnten Jahrhundert*, p. 5).

[45] "Wer mit seiner Kirchenordnung nicht friedlich sei, solle in Academia Julia nicht geduldet werden. Es sei besser dieselbe führe zum Teufel, als dass sie Kirche und Schulen veruntreuten und befleckten . . . ." (Emil Reicke, *Der Gelehrte in der deutschen Vergangenheit*, p. 102).

quired to accept the religion of the university.[46] That only Catholics frequented Catholic institutions is a fair assumption, although I found only that a *confessio fidei* was required for Würzburg. The greatest liberality prevailed at the Calvinistic University of Heidelberg. Here the professors had only to take oath that they accepted the word of God. Its statutes of 1672 declare that the professors must either abstain from entering into religious controversies or, if involved in them, develop in their lectures both sides with historical reasons.[47] To add an example of religious intolerance: Professor Weigel (1679) was forced formally to revoke his demonstration of the mystery of the Trinity from principles of arithmetic.[48] A case of intolerance of new philosophic views, closely related to religious intolerance, may be noted here. In Freiburg it was forbidden to refer to Peter Ramus except for refutation, and it was decreed that "no copy of his book must be found in the hands of a student."[49]

There was according to Eulenburg a distinct line of demarcation between the type of work in Protestant and Catholic universities. The Protestant universities laid emphasis on theology and law; they almost invariably had a faculty of medicine. The Catholic very often lacked the law and medical faculties, and persevered mostly along lines of the faculty of philosophy.[50]

[46] Tholuck, *op. cit.*, p. 9. How essential an element this acceptance was is best shown from the fact that in the matriculation tables *non juravit* is noted against the name of any who had omitted to give the oath to the established creed (Eulenburg, *op. cit.*, p. 96).

[47] The university even asked the Jew, Spinoza, to take the chair of philosophy, "trusting he would not abuse the privilege of philosophy for the overthrow of accepted religion." Spinoza refused, as he "did not know within what bounds freedom of philosophizing should be restricted so as not to interfere with public religion" (Tholuck, *op. cit.*, p. 9).

[48] *Ibid.*, p. 7.

[49] Dr. Heinrich Schreiber, *Geschichte der Albert-Ludwigs-Universität zu Freiburg*, II, 135.

[50] Eulenburg, *op. cit.*, p. 200, has the following data for the relative attendance of the four faculties:

1621–1700 for Strassburg, typical as the most progressive Protestant univer-

SCIENCE IN THE UNIVERSITIES

But they had in common many features which seem today most objectionable; for instance, the insufficiency of professorial staff. The average number of professors in the larger universities for the four faculties was, according to Eulenburg, sixteen. In Heidelberg (1558) a highly exceptional state existed, as there were three professors in theology, four in law, two in medicine, seven in philosophy, allowing one for Greek, Latin, logic, ethics, physics, and two for Hebrew and history.[51] Only the largest universities could afford so much specialization. Many of the Protestant foundations were so utterly poor that there the most incredible combination of professorial functions was made, for instance the same man teaching mathematics and medicine.[52] In Leipzig every professor read half a year, and the subject was decided by lot.[53]

Reading Aristotle and holding disputations were the main features of all instruction. The *Methodica facultatis artium* (1616) in Freiburg proscribed "omnia ex sententia Aristotelis et Peripateticorum doceantur."[54] In Erfurt there were, for instance in 1634, four disputations a year, in both theology and in medicine; monthly disputations in law, bi-weekly in philosophy.[55]

In this rather cheerless picture of German university life there are some instances of progress which deserve special attention, though Wegele very appropriately calls it progress, "in

---

sity: 20 per cent theology, 39 per cent law, 5 per cent medicine, 36 per cent philosophy.

1621–1700 for Würzburg, progressive Catholic university: 4 per cent theology, 7 per cent law, 5 per cent medicine, 51 per cent philosophy, 24 per cent rhetoric.

1621–1700 for Freiburg (Jesuit): 22 per cent theology, 21 per cent law, 2 per cent medicine, 55 per cent philosophy.

The astonishingly small number of students of medicine, the faculty where natural sciences would be cultivated, should specially be noted.

[51] *Ibid.*, p. 239.   [52] *Ibid.*, p. 251.   [53] *Ibid.*, p. 237.
[54] Schreiber, *op. cit.*, II, 133.
[55] Otto Bock, *Die Reform der Erfurter Universität*, p. 77.

äusserst gemässigtem Tempo."[56] As to establishment of scientific instruction, we find that often the professor of medicine taught physics, which seems an advance, from the point of view of physics.[57] In 1609 we hear of a special professor of chemistry in Marburg—Johann Hartmann; the first evidence of such a professorship in Europe.[58] In 1660 we know that Professors Gaspard Schott and Athanasius Kircher gave something like a course in experimental physics at Würzburg, for they showed the Guericke experiments,[59] but neither of them were modern physicists.[60] Papin was for a while professor of physics in Marburg, 1695.[60]

The great and conspicuous instance of the cultivation of science is the strangely progressive University of Altdorf, a creation of the city of Nürnberg.[61] Here Professor Christian Sturm, who is already known to us as the founder of the Collegium Curiosum sive Experimentale, gave to his students in 1683 a course of experimental physics, comprising the experiments of the Cimento, which he had witnessed in Florence. Here Trew, a most progressive professor of astronomy, had a tower remodeled into an observatory, and had prosperous burghers supply it with

[56] Franz X. Wegele, *Geschichte der Universität Würzburg*, p. 346.

[57] Schreiber, *op. cit.*, II, 66. In Freiburg, physics was intrusted to professors of medicine, *propter professionum affinitatem et tenues facultates academiae;* and O. H. Arnoldt, *Historie der Königsbergischen Universität*, p. 208, says: "of three members of the medical faculty, the first should read chemica in addition to practica, and the second should read physics as long as the philosophical faculty has no special professor of physics, but if they have one, he should read anatomy in winter and botany in the summer."

[58] J. C. Poggendorff, *Biographisch-literarisches Handwörterbuch*.

[59] W. Lexis, *Das Unterrichtswesen in Deutschen Reich*. Vol. I: *Die Universitäten im Deutschen Reich*, p. 407.

[60] Rosenberger, *op. cit.*, pp. 120 ff. Schott in 1657 believed in the *horror vacui*, and Rosenberger calls Kircher an old-fashioned physicist who simply used some experimental methods.

[61] S. Günther, "Die mathematischen- und Naturwissenschaften an der nürnbergischen Universität Altdorf.," *Mitteilungen des Vereins für Geschichte der Stadt Nürnberg*, Heft 3, p. 9.

instruments.[62] Here, most remarkable of all, in 1682 Moritz Hoffman built, with the funds of the university, a well-equipped chemical laboratory, the first chemical university laboratory of the world,[63] a great contrast to Erfurt, for instance, where the professor announced a "Promptuarium Pharmacopoesios Chymicae in privato Vulcani domestici Laboratorio."[64]

Botany was somewhat more favored than chemistry or physics. Throughout the century there is frequent mention of the establishment of botanical gardens, so that most universities seem to have been supplied with them.[65] We also hear occasionally of botanical excursions.[66]

In anatomical study the necessity of dissection was recognized in many universities during the century, so, for instance, anatomical theaters were erected in Jena (1629),[67] in Freiburg (1620),[68] in Altdorf (1637), and in Würzburg (1675).[69] Of the

[62] *Ibid.*, pp. 6 f.

[63] See illustration No. 107 in Herman Peters, *Der Arzt und die Heilkunst in deutscher Vergangenheit*. I add the description as given by Günther, *op. cit.*, pp. 10 ff.: "Among its furnaces the chief are: first, to the right the *Furnus docimasticus* for the testing of metals; next to this the *Balneum Mariae sive Vaporosum*, serviceable for all sorts of distillations; next, a great open fire with bellows for melting metals; then still another furnace, *Piger Henricus*, with two sand baths on each side, which could both be heated at the same time; next, on a table, a copper stove, *Lampas Philosophiae*, which was undoubtedly used to search for the great mystery of the philosopher's stone. Between the third and fourth windows to the left there was a special furnace for operations which required a particularly strong fire; and then there was a furnace for distillation in an open fire; then a *Furnus sublimatoricus* for the preparation of volatile substances; finally, a great open fire. Then in front two more furnaces, one of clay with a sand bath for common distillations, the other of iron for the boiling of juices or other liquids. This was movable. In the middle of the room was a long table and a chair for purposes of instruction, in which the *Chemiae Doctor* dealt at the same time with theory and practice."

[64] Bock, *op. cit.*, p. 71.

[65] Günther, *op. cit.*, p. 5. Günther mentions, however, that the University of Ingolstadt refused the establishment of a botanical garden.

[66] Schreiber, *op. cit.*, II, 152. [68] Schreiber, *op. cit.*, II, 147.
[67] Lexis, *op. cit.*, p. 579. [69] Wegele, *op. cit.*, p. 385.

anatomical theater in Altdorf pictures are extant.[70] In Würzburg the appointment of a special demonstrator (1674) in anatomy indicates that anatomical studies commenced to flourish there.[71] Indeed, we hear of a popular insurrection in 1661 against *rolfinking* the body of a criminal—a word meaning "dissecting" and coined from the name of Rolfink, the Würzburg professor of anatomy.[72] It would in this connection be interesting to know how much practical experience was required for the medical degree. But for this I find only one reference in the case of Erfurt,[73] where the student had to submit to a practical examination, "in theatro anatomico, horto botanico, laboratorio chymico et officina cum chirurgicatum pharmaco."

We have so far followed the *äusserst gemässigt* progress of the seventeenth century in the universities only as regards their attention to the natural sciences. We shall turn now to the other line along which progress was made in Germany, namely, the establishment of the vernacular. For the decisive move, the burning of the bull of this reformation, occurred in the seventeenth century.[74] The Luther of this movement is one of the most interesting personalities of German history, and the hero of German university reform, Christian Thomasius.

He was professor of law in the very conservative University

[70] Günther, *op. cit.*, pp. 14 f., gives the following description: "There was a dissection table to which water was conducted from the yard; the benches were arranged in the form of an amphitheatre; skeletons stood about, and drawings made by the Altdorf Professor Hoffman. Besides [and this is most interesting], there was a great number of animal skeletons, apparently used for comparative anatomy."

[71] Wegele, *op. cit.*, p. 405.

[72] Puschmann, *op. cit.*, pp. 14 f.

[73] Bock, *op. cit.*, p. 71.

[74] A solitary instance of a university lecture in German dates back to 1527, when Paracelsus in Basel addressed his students in the vernacular. Eulenburg, *op. cit.*, p. 110, records an instance of the use of the vernacular at the University of Rostock, where lectures were held in Low German; and in Königsberg (Arnoldt, *op. cit.*, p. 83) lectures on mensuration were given in the vernacular.

SCIENCE IN THE UNIVERSITIES

of Leipzig, when in 1679 he nailed on the blackboard, "never before so desecrated," a German announcement of his lecture. Great uproar ensued—such a thing had never happened; it seemed as if they wanted in solemn procession to purify the board with holy water.[75] Breaking away from all precedent, Thomasius donned the modern garb, lectured in the German language on so unlearned a topic as "Prejudices,"[76] and even dared to offend the faculty by handing to them a book written in the vulgar tongue which they of course rejected.[77]

In his essay on *Imitation of the French,* Thomasius claimed that "not the useless knowledge of Latin, but usefulness in life is the test of man's wisdom. The French, who have adopted their own language, should be imitated. Thus, women would not be excluded from all education."[78]

He never ceased his defiance of Latin. In 1687 he wrote:

It is a principle of politics that the ruler must accustom his subjects to the language of the ruler. This principle the Pope has adopted and requires all priests to use the Latin language as a sign of their subjection to him. Ever since the time of Charlemagne this superstition has been introduced into universities, and in order that they should be withdrawn from the supremacy of temporal rulers, all professors and students were included in Holy Orders. Thus Latin became the language of the learned only because it was the language of the parsons [*Pfaffen*].[79]

And again in his *Vernunftlehre* Thomasius wrote:

Foreign languages are of use in order to understand what has been written in them; but in those things which are realized through the intelligence which is innate in all nations, knowledge of foreign languages is not at all necessary. Worldly wisdom is so easy a matter, that it can be understood by all people; Greek philosophers did not write in Hebrew but in their mother tongue......[80]

[75] Prutz, *op. cit.,* pp. 306 f. (excellent account).
[76] Ludwig Salomon, *Geschichte des deutschen Zeitungswesens,* I, 92.
[77] H. Dernburg, *Thomasius und die Stiftung der Universität Halle,* p. 13.
[78] Christian Thomasius, *Vom Nachahmung der Franzosen.* "Deutsche Litteratur denkmale des 18. und 19. Jahrhunderts," No. 51, p. 25.
[79] Hodermann quoting Thomasius. [80] Salomon, *op. cit.,* I, 92–97.

Not satisfied with the blow he dealt scholasticism by striking at Latin, Thomasius started in 1688 the publication of a periodical *Monatsgespräche. Scherz- und Ernsthafter Vernünftiger und Einfältiger Gedanken, über allerhand lustige und nützliche Bücker und Fragen.* Its purpose was to ridicule the entire university system.[81] When, after several issues, Thomasius was forced to leave Leipzig, he fled to the Elector (later King), Frederick of Prussia, who granted him *venia legendi* in Halle. This led in 1690 to the foundation of the University of Halle, the first reformed university of Germany, which became the model of all other university reform. In the first lecture in Halle, which of course was in German, Thomasius characteristically exclaimed: "We are not bound to Aristotle, we shall not be accused of *lèse majesté* even if we make fun of the king of philosophers, and philosopher of kings."

So, at the end of the seventeenth century, there existed in Germany, one center at least, at the University of Halle, where the vernacular and anti-Aristotelianism were cherished, and one,

[81] The theme of the first issue and in a lesser degree of all the following was an attack against the learned pedantry and the unreality of university life. We have here a parallel to the *Epistulae virorum obscurorum* in seventeenth-century garb. In a coach traveling from Frankfort to Leipzig are four men: Augustin, a French traveler; Benedict, a learned man; Christoph, the practical business man voicing Thomasius' ideas; and David, a schoolman, the soul of pedantry. The conversation starts with a discussion of a book of Abraham a Sancta Clara. David disapproving of such books written in the vernacular, "for their loose morality," Christoph hints that this scrupluous man would not refrain from reading "purissimam impurissimi scriptoris Martialis Latinitatem." Then he pours sarcasm on this scholastic attitude, and the hairsplitting over whether King David took coffee, or Dido smoked tobacco with Aeneas. Only, he says, in Holland, where there is liberty of the press, can real learning flourish. As the discussion turns on the *Acta eruditorum* the coach is upturned. The March issue made fun of the four faculties and therefore the professors wanted Thomasius discharged. Then the April paper turned with such bitterness against Aristotle and the professors that nobody dared to reply. In the December number, Thomasius directed his arrows against the interference and narrowness of a Danish court preacher who declared Lutheranism the one and only sound religion (Thomasius was Lutheran). This caused him to lose his position, and the license to continue to publish his journal, and exposed him to imprisonment (*ibid.*).

at the University of Altdorf, where pure science had found a place of encouragement.

ENGLAND

A review of the English universities during the seventeenth century presents to us almost insurmountable difficulties, because in all accounts of their activities stress is laid almost exclusively upon the political and religious features of their history. Of course the universities were in such close and direct connection with the political disturbances and endless religious wranglings of the Stuart and Cromwell period that this attitude of the historian is perhaps natural, and is in itself an important commentary upon the situation.[82] On the other hand, many criticisms of the university system were published in England during the century, and these not only give us information in regard to the alleged defects of the system but also make clear that there were men who recognized them and tried to concentrate public attention upon their reformation.

England in 1600 had its two great colleges, Oxford and Cambridge;[83] Scotland had four universities, Aberdeen, St. Andrews, Glasgow, Edinburgh; Ireland had one university, that of Dublin. The general features of these English colleges in 1600 as established by the Elizabethan statutes in 1570[84] were such as prevailed generally throughout Europe.[85]

[82] I find the following statement in the *Register of the University of Oxford* (*1571–1622*), II, Part I, 227: "The question presents itself: How did the old course give way to the present one? Its stages are without doubt recorded in successive codes of statutes. It seems strange it should still await an historian."

[83] Gresham College, which has been discussed above, as well as the Royal College of Physicians, occupied a peculiar position. The latter was throughout the century the home and advocate of truly scientific methods, and produced, in Harvey, Willis, and Sydenham, men who took a leading place in their science throughout Europe.

[84] *Register of the University of Oxford*, II, Part I, 107

[85] The main distinction was that nowhere else, perhaps, was the ecclesiastic element so very prominent a feature. During the religious controversies of the seventeenth century, the universities were constantly drawn into and not infrequently became the center of the wrangling.

Gauging the prevalent interest in science by the faculty of medicine, which may be called the barometer of scientific interest, we find, according to the *Register of Oxford,* that it was at the lowest ebb conceivable;[86] it was no more than the cerement of dead learning, and had lost all touch with professional study. Only a small number of students attached themselves to this faculty; indeed, it would have become extinct but for the existence of the Regius professorship, endowed by Henry VIII, and of fellowships in medicine in the various colleges.[86] It was only after 1626 that a slight knowledge of anatomy was required from the medical students at Oxford,[87] and even then frequent dispensations are registered.[88] The level of medical instruction at Oxford[89] may be seen from the titles of a few disputations:

1605. Incantatio non valet ad curam morbi.
1608. An vita hominis sine respiratione consistere possit? [Affirm.]
1608. An omnes corporis partes sanguine nutriantur? [Neg.]
1611. Aegri opinio de medico facit ad salutem.
1611. Medicamenta non sunt cibis conmiscenda.
1613. An liceat morbum morbo curare?
1615. An ciborum varietas sit praeferenda cibo simplici? [Neg.]
1618. An mulieres a melancholia magis vexentur quam viri? [Affirm.][90]

Following the development of the universities mainly in their relations to natural sciences, we find the first significant innovation in Oxford in 1619. Savile, who had lectured gratuitously at Oxford on Greek and geometry, endowed two professorships, one of geometry and one of astronomy, enjoining upon the lec-

[86] *Register of the University of Oxford,* II, Part I, 123.

[87] Perhaps owing to the establishment of the Tomlins professorship of anatomy, 1624.

[88] *Register of the University of Oxford,* Part I, p. 123.

[89] As to medical instruction at Cambridge, I have no definite information, but assume that a similar state prevailed.

[90] *Register of the University of Oxford,* II, Part I, 191.

turer to teach optics and to mention the Copernican system.[91] A similar endowment was made by Savile's son-in-law, William Sedley,[92] in founding a lectureship in natural philosophy at Oxford (1621).[93]

In 1636, Archbishop Laud, as chancellor of Oxford, promulgated the Laudian statutes, which were destined to remain in force at Oxford until 1854. He threw his influence upon the side of conservatism,[94] and the innovations his statutes introduced were distinctly reactionary.

The importance of dialectics and the authority of Aristotle were to be strenuously inculcated, it being especially enjoined that . . . . Regent should . . . . deliver an address expressly designed to vindicate the above features. . . . . In the B.A. course the subjects were to include grammar, rhetoric, Aristotle's ethics, politics, and economics; logic, moral philosophy, geometry and Greek. In the M.A. course, more geometry, more Greek, together with astronomy, metaphysics, natural philosophy and Hebrew. All students admitted to a degree should give evidence of possessing a good command of correct colloquial Latin.[95]

[91] The holders of the Savilian professorships during the century were: in geometry—Briggs, Wallis, Halley; in astronomy—Bainbridge, Greaves, Seth Ward, Wren, E. Bernard, David Gregory, John Keil.

[92] Sedleian professors of natural philosophy during the century were Lapworth, Edwards, Crosse, Willis, Millington.

[93] Mullinger, *op. cit.*, III, 84. For Cambridge there falls into this period the founding of a lectureship which, although not in science, is of interest to us on account of its "lay character." Lord Brooks endowed a lectureship in history, "open to foreigners but not any one in holie orders, because this realm affords many preferments for divines, few or none to professors of profane learning, the use and application whereof to the practice of life is the maine end and scope of this foundation."

[94] *Ibid.*, p. 91: "His views strongly resembled, ... were in some respects almost identical with the theory of education advocated by the Jesuits. He was desirous to widen the field of knowledge of the universities, to render their treatment of the ancient trivium and quadrivium more intelligent and thorough, and more especially to give to philology an importance and prominence far greater than it had as yet attained. . . . . But here like the Jesuits, he halted."

[95] *Ibid.*, p. 135.

These Laudian statutes were adopted in the University of Dublin in 1637.[96]

The first class studied Logic, especially the *Isagoge* of Porphyry, which was required to be read over twice at least in the year. The lecturer of the second class explained some part of Aristotle's *Organon* as briefly as possible, not allowing himself to wander from the context into commentaries upon the text. The lecturer of the Junior Sophisters read with his class some portions of the *Physics* of Aristotle. The lecturer of the Senior Sophister [fourth class] took up the *Metaphysics* of Aristotle, except in Lent Term when he read with his class the *Nichomachean Ethics*.[97]

Similar regulations I find for the University of Glasgow.[98] The third year was given to Aristotle's *Logic,* the fourth year to his *Ethics, Politics* and *Economics, Metaphysics,* and to the first two books of his *Physics;* the fifth year to the remaining books of the *Physics.*[99] These regulations remained in force until 1688.

Latin was used exclusively. In 1649 "the Committee for regulating Universities" decreed that only Latin *or Greek*(!) was to be used in colloquial discourse among students. "No other language was to be spoken by any fellow, scholar or student whatsoever."[100] The fact that all textbooks of mathematics and physics throughout the seventeenth century were in Latin is a convincing proof of the persistence of this custom.

[96] J. W. Stubbes, *The History of the University of Dublin (1591–1800),* p. 139: "In the Statutes prepared by Laud, very express directions are given with respect to the details of the education to be imparted. These appear to be in harmony with the teaching which was prevalent from the first foundation of the College; and although they were obsolete before they were finally repealed, they left an impress upon the studies pursued in the College for more than a century."

[97] *Ibid.,* p. 140.

[98] James Coutts, *The History of the University of Glasgow (1451–1909),* p. 139: "It is remarkable how much of the curriculum in 1641 is still dominated by Aristotle."

[99] *Ibid.,* pp. 101–11. The reading of a compendium on anatomy was, however, required in the fifth class at Glasgow.

[100] Mullinger, *op. cit.,* III, 368.

SCIENCE IN THE UNIVERSITIES

The most enlightening commentary on the condition of English universities in the first quarter of the century is Francis Bacon's criticism of the system. As he was a graduate of Cambridge, his views on universities in general were formed from conditions prevailing in his own Alma Mater. Bacon found that the whole university system needed readjustment.[101] He directed his attacks again and again against the professors who saw their task merely in commenting upon the accepted knowledge,[102] and who felt it their duty to impress their students with its finality[103] rather than lead them along new lines of inquiry.[104] He explained that sciences could not advance because they had not been considered as special branches of inquiry,[105] because in-

[101] Bacon, *Of the Advancement of Learning:* "As .... usages and orders of the universities were derived from more obscure times, it is the more requisite they be re-examined."

[102] *Ibid.:* "For whereas the more constant and devoted kind of professor of any science ought to make some addition to their science, they convert their labors .... to be a profound interpreter or commenter, a methodical compounder; or abridger."

Bacon, *Novum organum*, Book I, Aph. XC: "Again in the customs and institutions of schools, academies, colleges and similar bodies destined for the abode of learned men and the cultivation of learning, everything is found adverse to the progress of science. .... For the studies of men in these places are confined, and, as it were, imprisoned in the writings of certain authors, from whom if any man dissent, he is straightway arraigned as a turbulent person or innovator."

[103] *Ibid.*, Book I, Aph. LXXXVI: "They set [their truths] forth with such ambition and parade .... as if they were complete in all parts and finished."

[104] Bacon, Author's Preface, *The Great Instauration:* "I .... seek .... to lead them to things themselves .... that they may see for themselves what they have .... what they can add .... to the common stock."

[105] Bacon, *Of the Advancement of Learning:* "Among so many colleges in Europe, I find it strange that they are all dedicated to professions, and none left free to arts and sciences at large."

Bacon, *Novum organum*, Book I, Aph. LXXX: "Natural philosophy even among those who have attended to it, has scarcely ever possessed a disengaged whole man; but it has been merely a passage and bridge to something else. So this great mother of sciences has with strange indignity been degraded to the office of a servant; having to attend on the business of medicine or mathematics, or like-

struments had not been contrived,[106] and the method of experimentation had been scorned by university men.[107] Yet Bacon dedicated both his *Novum organum* and *De augmentis* to Cambridge University as "its son and nursling, repaying his indebtedness as far as was in his power, inasmuch as he imbibed his first draught of knowledge at its sources."[108] It may be questioned whether this was a sign of devotion, or an attempt to impress his ideas more forcibly on his Alma Mater.[109]

Indeed, the decades during which Bacon's works were written mark the lowest level Cambridge reached. Ball says that while up to 1600 there was a succession of mathematicians at

---

wise to wash and imbue youthful and unripe wits with a sort of first dye in order that they may be the fitter to receive another afterward. . . . . For want of this, astronomy, optics . . . . mechanical arts, medicine . . . . altogether lack profoundness and merely glide along the surface."

[106] Bacon, *Of the Advancement of Learning:* "Another defect I note wherein I shall need some alchemist to help me, who call upon men to sell their books and build furnaces. . . . . Unto . . . . study of many sciences, specially natural philosophy and physics, books be not only the instruments; wherein there . . . . will be hardly any main proficience in the disclosing of nature except there be some allowance for expenses about experiments, appertaining to Vulcan or Daedalus, furnace or engine . . . . you must allow the spials or intelligencers of nature to bring in their bills. . . . ."

[107] Bacon, *Novum organum*, Book I, Aph. LXXXIII: "[There exists] an opinion . . . . that the dignity of the human mind is impaired by . . . . intercourse with experiments . . . . especially as they are laborious to search, ignoble to meditate, harsh to deliver, illiberal to practise, infinite in number and minute in subtility. So that it . . . . comes to this . . . . that . . . . experience . . . . is rejected with disdain."

Bacon, *Of the Advancement of Learning:* "It is esteemed a kind of dishonor unto learning to descend to inquiry or meditation upon matters mechanical, except they be such as may be thought secrets, rarities and special subtilties."

[108] Mullinger, *op. cit.*, III, 67.

[109] This interpretation would be borne out by the fact that in Bacon's will appeared the design of founding at Cambridge and Oxford a lectureship in natural philosophy, "with science in general thereunto belonging, the professor to be appointed without difference, whether a stranger but not professed in divinity, law or physic." The paucity of the funds he left prevented, however, the establishment of these endowments.

SCIENCE IN THE UNIVERSITIES

Cambridge, the next thirty years form a blank in the history of that science.[110] Students found very discouraging conditions. Horrox, who wanted to study astronomy in 1633, had to do so from books as this subject was not offered at Cambridge. Wallis left the following account of his studies (1635):

> Mathematics were scarce looked upon as Academic studies, but rather Mechanical, and among more than two hundred students in our college, I do not know of any two (perhaps not any) who had more of mathematics than I (if so much) which was then but little, and but very few in that whole University. For the study of mathematics was at that time more cultivated in London than in the Universities.[111]

Further it is of interest to note that after Wallis, through his own efforts and talent, had acquired his great knowledge of mathematics, he had to leave Cambridge—because that study had died out there[111]—and accept the Savilian professorship of geometry at Oxford (1649–1702). His colleague there in astronomy was another Cambridge man, Seth Ward, who—as was said above—had been expelled from his Alma Mater for refusing to subscribe to the "League and Covenant."[112]

Thus Oxford gained what Cambridge lost. Indeed, at Oxford, in the middle of the century a remarkable set of men held professorships, most of whom we mentioned before as belonging to the Invisible College:[113] Wallis, Ward, John Wilkins (warden of Wadham), Willis, Rooke, Wren, and others. We have seen these men experimenting in their own quarters, at their own expense. It is important for us to decide whether the private investigations of these scientists in collaboration with men like Boyle, unaffiliated with the university, are to be credited to the university or not. That their efforts were really disconnected with Oxford University is shown by the fact that when these in-

[110] W. W. Ball, *A short account of the history of Mathematics*, p. 35.
[111] Mullinger, *op. cit.*, III, 462.
[112] Ball, *op. cit.*, p. 37.
[113] See above, chap. iv.

dividual students withdrew from Oxford, the cause of science was utterly deserted, and Uffenbach in 1710, visiting the "chemical laboratory fitted for the original Royal Society, finds the stoves in fair condition, but everything else in disorder and dirt."[114]

In the middle of the century there was a flood of criticism leveled against the university system, some of which are truly interesting and bear directly upon the query we have raised. There was John Hall's *Humble Motion to the Parliament of England concerning the Advancement of Learning and Reformation of Universities* (1649),[115] in which is found the following significant statement:

> I have ever expected from an university, that though all men cannot learne all things, yet they should be able to teach all things to all men; and be able either to attract knowing men from abroad out of their owne wealth, or at least be able to make an exchange. But how far short come we of this, though I acknowledge some differences between our universities? We have hardly professours for the three principall faculties, and these but lazily read,—and carelessly followed. Where have we anything to do with Chimistry, which hath snatcht the Keyes of Nature from the other sects of philosophy by her multiplied experiences? Where have we constant reading upon either quick or dead anatomies, or occular demonstrations of herbes? Where any manual demonstrations of Mathematicall theorems or instruments? Where a promotion of their experiences which if right carried on, would multiply even to astonishment?[116]

Then there was the interesting invective against Cambridge of William Dell, master of Caius College (1649)[117] and the fa-

[114] Wordsworth, *op. cit.*, p. 176.

[115] Mullinger, *op. cit.*, III, 371: "John Hall's indictment is certainly entitled to be considered . . . . valuable in respect of precision and as giving forcible utterance to convictions already lurking in the mind of not a few who had neither the courage nor ability to set them forth with equal force and plainness."

[116] Hall, *op. cit.*, pp. 4 f.

[117] Dell appears as one of the earliest English writers to insist on the education of the people as the foremost duty of the state, as distinguished from the church. In schools he advised that there should be a more extended range of

mous criticism in Hobbes's *Leviathan*.[118] The pamphlet, however, which acquired the most elaborate reputation was Webster's *Academiarum examen,* or the "Examination of Academies."[119]

Webster directs what he has to say as a critic chiefly to the existing "customs and methods" of the Schools with their scholastic exercises, urging, as a serious objection, that in all such exercises "they make use of the Latin tongue .... whereby the way to attain Knowledge is made more difficult and the time more tedious, and so we almost become strangers to our own mother tongue." The stress of his criticism, however, is concerned with the defects of the existing curriculum rather than its abuses, and here the justice of his comments is so obvious, that it seems difficult to understand how more than another century was yet to pass away, before his suggestions were carried, even partially, into effect. He dwells upon the desirability and excellence of physical studies; he deplores the neglect of mathematics; the "sloathfulness and negligence of the professors and artists," as a body, describing them as ignorant "that their scrutiny should be through the whole theatre of nature," and that "their only study and labour ought to be to acquire and find out salves for every sore and medicines for every malady, and not to be enchained with the formal prescriptions of schools, Halls, colleges, or masters." Then he turns to extol that great discovery of Harvey, "our never sufficiently honoured countryman," and expresses his regret that it has not been more generally utilized. He dwells with like em-

---

subjects; pupils should be taught first to read their native tongues. In universities he wanted mathematics to be held in good esteem, and physics and law to be studied. He thought it would be advantageous to the people to have in every great town, as in London, York, etc., at least one university or college which should be supported by the state (Mullinger, *op. cit.,* III, 454 ff.).

[118] Hobbes, sent abroad, saw the worthlessness of much of the scholastic philosophy he learned in Oxford, and in the *Leviathan* assailed the system of universities as originally founded for the support of papal *vs.* civil authority, and as working social mischief by their adherence to the old learning (*ibid.,* p. 433).

[119] "Wherein is discussed and examined the Matter, Method and Customes of Academick and Scholastic Learning, and the insufficiency thereof discovered and laid open; also some expedients proposed for the reforming of Schools and the perfecting and promoting of all kinds of Science. Offered to judgment of all those that love the proficience of Arts and Sciences and the Advancement of Learning" (London, 1654). [Complete title of above work.]

phasis on the merits of Gilbert's treatise, *De magneto*. "What shall I say," he asks, "of the atomical learning revived by that noble and indefatigable person, Renatus Des Cartes?"[120]

The reply to this attack is, for our study, of greatest importance, because it was written by Bishop John Wilkins and Seth Ward and is, as it were, a declaration of faith from the group of scientists we met before as the Invisible College, or Oxford Society.[121] Webster's attack was regarded as utter folly by these men. Wilkins undertook to reply annonymously to Webster's insinuation that Aristotle still was idolized at the universities,

whereas those who understand those places, do know that there is not to be wished a more general liberty in point of judgment or debate than what is here allowed. So that there is scarce any hypothesis, which hath been formerly or lately entertained of judicious men, and seems to have in it any clearness or consistency, but hath here its strenuous assertours, as the atomical and magneticall in philosophy, the Copernican in astronomy, etc. And though we do very much honour Aristotle, yet are we so farre from being tyed up to his opinions, that persons of all conditions amongst us take liberty to dissent from him, and to declare against him, according as any contrary evidence doth engage them, being ready to follow the Banner of Truth by whomsoever it shall be lifted up.[122]

Ward replied to the more specific attack, that sciences were not fostered at the two universities—with the bitterness of one who felt that his own life's work was being overlooked. He objected strenuously to pamphleteers, who under pretext of giving effect to the teachings of Bacon dared to urge that

instead of verball exercises we should set upon experiments and observations, that we should lay aside our Disputations and Declamations and Public Lectures and betake ourselves to Agriculture, Mechanicks, Chymystry, and the like. . . . . Our Academies are of a more general and comprehensive institution, and as there is a provision here made that whosoever

---

[120] Quoted from Mullinger, *op. cit.*, III, 458.

[121] *Vindicae Academiarum containing some briefe animadversions upon Mr. Webster's Book stiled, The Examination of Academies. Together with an Appendix concerning what M. Hobbes and M. Dell have published on this Argument.*

[122] *Ibid.*, p. 2.

will be excellent in any kind, in any art, science or language, may here receive assistance, and be led by the hand, till he come to be excellent: so is there provision likewise that men be not forced into particular ways but may receive an institution variously answerable to their genius and design.[123]

Yet Ward admitted that

it cannot be denied but this [experimentation] is the way and the only way to perfect Natural Philosophy and Medicine; so that whosoever intend to professe the one or the other, are to take that course and I have not neglected occasionally to tell the world that this way is persued amongst us.[123]

He would thus imply that regular courses in science were given at Oxford.[124] If so, we have failed to find any trace of them. It may be assumed that the excellent work of these pioneers was *extra collegium* and does not affect the justice of Webster's criticism as subsequent years amply showed.[125]

But we must leave these criticisms, where our sympathy seems inevitably on the side of the critics, to say a few words about the state of science at the universities in the second half of the century. At Oxford, we have seen that in 1659 all chemical instruction was given by Peter Stahl, the Strassburg chemist, whom Boyle had induced to settle in his house, without any affiliation with the university.[126] For 1671 I find a letter showing dissatisfaction at the little aid Oxford gave to science in spite of the interest of the students:

[123] *Ibid.*, pp. 49 f.

[124] Mullinger, *op. cit.*, III, 466, in a footnote adjoins: "We accordingly here have it on unimpeachable authority of a professor of the University of Oxford; 1654, that at that time any student desirous of specializing in natural science [i.e., medicine, chemistry, or mineralogy] with a view to a professional career, was allowed to do so."

[125] There was joined to the reply to Webster an answer to the criticism of Hobbes which forms but a small chapter in the protracted controversy of Ward and his friends with the author of the *Leviathan;* also a venomous reply to Dell's charge that Oxford and Cambridge were monopolizing education, in which Ward asserted absurdly enough that such a statement implied an unwarranted criticism of the sovereign magistracy of the nation (*Vindiciae,* p. 63).

[126] Clark, *op. cit.*, I, 290.

The mathematics [in Oxford] I own are but just enough to admire Dr. Wallis here . . . . and the remainder of my study is litteral, and beside the fame and regard of this age, and inferior in the nature of things to real learning. . . . . And this I must always affirm for the honour of my mother, the University of Oxford: if her children had the good utensils which adorn colleges of Jesuits abroad, the world would not long want good proof of their ingenuity. Patrons and tools are rather wanting, than willing and fit workmen. We lack a corporation, a set of grinders of glasses, instrument makers, operators . . . . that experiments may be well managed in this place which otherwise, by reason of our living together and our freedom from intricacies and vexation of the world, is most convenient for such a design.[127]

Another evidence that Oxford students were interested but that the university supplied no aid is the following letter of Dr. Beal to Boyle (November 27, 1671):

At my request a young Oxonian prepared me a list of fit capable and hopeful persons addicted to the design of the Royal Society, and willing to entertain correspondences and to assist them. They seemed to me by their qualifications and number very considerable, some in every college and every hall, ... There are excellent professors, some lecturers and very many students of useful arts amongst them. And in time they may have their meetings in some of their public schools.[128]

Yet we are told that the Oxford pulpit declared Boyle's researches were destroying religion and his experiments undermining the universities.[129] Similarly, Wilkins was preached at from the university pulpit as a "mere moral man without power of godliness."[130]

But if in 1671 Oxford had not much science, it had at least some of the odium attached by conservatives to the thought of innovation. Wordsworth relates as follows:

[1671] Dr. John Eachard, in "Some observations in answer to enquiry into grounds of occasion of Contempt of Clergy" [pp. 142–47], gives an

[127] Rigaud, *op. cit.*, I, 158 (Dr. E. Bernard writing to Collins).

[128] Thomas Birch, *The Works of Robert Boyle*, V, 498.

[129] Andrew D. White, *A History of the Warfare of Science with Theology in Christendom*, I, 405.

[130] Wells, *op. cit.*

SCIENCE IN THE UNIVERSITIES

amusing sketch of a pert young academical sciolist: "And in the first place comes rattling home from the university the pert Sophomore with his atoms and globuli; as full of defiance of all country parsons, let them never be so learned and prudent, and as confident and magisterial as if he had been prosecutor at the first Council at Nice, ... They are all so sottish and stupid as not to sell all their libraries and send presently away for a whole wagon full of new Philosophy. I'll tell you, Sir, says one of these small whiflers—the University is strangely altered since you were there, we are grown strangely inquisitive and ingenious. I pray, Sir, how went the business of motion in your days? We hold it all now to be violent. . . . ."

Then follows a slash at younger members of Gresham College who ask "to what purpose is it to preach to people and go about to save them, without a telescope or glass for Fleas."[131]

In 1683 the Ashmolean Museum was furnished with a chemical laboratory—*officina chymica*—and Dr. Plot, its first curator, was ultimately appointed professor of chemistry. In an account of Worcester Hall I find that in 1698 courses in anatomy, chemistry, and botany were given there. The anatomist, however, was in his lectures to comment on the first verses of the twelfth chapter of Ecclesiastes in order to explain his anatomical teachings. The chemist gave four lectures on the principles of chemistry and twelve in experimental chemistry; the botanist four lectures in general botany, eight in practical botany, and was to take his hearers four times into the field.[132] On the whole, science seems little cultivated at Oxford in 1700.[133]

Cambridge seems to have had more of a reputation for "new philosophy" than Oxford. Here the ideas of her great son Bacon were in a measure cherished, so that Glanville regretted not having gone there.[134] It is a mark of progressiveness that the Cam-

---

[131] Wordsworth, *op. cit.*

[132] C. H. Daniel and W. R. Barker, *Oxford University, College Histories, Worcester*, p. 160.

[133] "Oxford," *Encyclopaedia of Education*: "It is probable that no University began its revival so late as the University of Oxford. That revival corresponds to the beginning of the 19th century."

[134] Fowler, *op. cit.*, p. 126.

bridge Platonists opened their doors widely to Cartesianism, while other universities, e.g., Glasgow, took the Jesuit attitude of abhorrence to such innovations.[135] But as late as 1662 Ray, the zoölogist, lost his position on account of objecting to the form of the oath.[136] The sudden rise of mathematics and the professorship of Newton give to Cambridge during the years 1660–1700 a significance in the history of science which is recognized more clearly today than it was by contemporaneous students there.

We saw how little astronomy and mathematics were cultivated in Cambridge in the first half of the century. At last in 1663, the Lucasian professorship was endowed with the condition that the holder might lecture on geometry, arithmetic, astronomy, geography, optics, statics, or other branches of mathematics, and that the fund was also to be used for the purchase of mathematical books or instruments.[137] Isaac Barrow (1630–77) was the first to hold this chair, and his lectures on mathematics and optics constitute the first significant effort of Cambridge in the realm of pure science. They failed to attract considerable audiences, and Barrow in 1669 resigned in favor of his pupil, Isaac Newton.[138] While it is needless to say what Newton's work in mathematics, optics, and celestial mechanics stands for, it is important to emphasize again two points: first, Newton[139] was exactly of the same type as Boyle and Hooke, a tireless experimenter without direct inspiration from his college, at first equally interested in chemical and alchemistic experiments, working at an apothecary shop, or in his garden or bedroom arranged for laboratory purposes; second, no word of the

[135] Mullinger, *op. cit.*, III, 495.

[136] Carus, *op. cit.*, p. 428.

[137] *Trusts, Statutes and Directions Affecting Professorships, Scholarships and Prizes and Other Endowments of the University of Cambridge*, p. 30.

[138] Ball, *op. cit.*, p. 48.

[139] David Brewster, *Life of Sir Isaac Newton*, p. 265.

SCIENCE IN THE UNIVERSITIES

revolutionary changes in optics contained in his lectures reached the world of science from the lecture-room of the Lucasian professor until he, upon Ward's advice, reported the matter to, and became a member of, the London Royal Society. Indeed, the fact that in spite of Newton's efforts no society for experimentation could be formed in Cambridge during those years shows most conclusively the lack of enthusiasm for science there.[140]

The story of how the Newtonian philosophy[141] took root in English universities is interesting as illustrating university conservatism. By an odd accident, just at the time when Newton's work was conceived—which was to deal the deathblow to Cartesian physics—Descartes' system, after long resistance on the part of the Aristotelian physicists, had come to be adopted; and, under a Cartesian garb, Newton's philosophy was to enter Cambridge.[142] In 1697 Dr. Samuel Clarke translated Rohault's *Physics*, the standard textbook of the Cartesian system, into Latin, and in the notes the translator gave by way of comment Newton's views, which were a refutation of the Cartesian text. The odd title of this composite book was *Jacobus Rohaultus physica Latine reddita et annotata ex, Js. Newtonii principiis*. Through this strategem Newton's ideas were forced upon the consideration of the instructors and tutors whose prejudice it was at once more important and more difficult to overcome than that of the professors.[143] Newton himself held the Lucasian professorship until 1702. Then he was succeeded by Professor

[140] Newton, referring to these efforts, wrote: "That which chiefly dasht the business was the want of persons willing to try experiments. The one we chiefly rel'd on refusing to concern himself in that kind. And I should be very ready to concur with any persons for promoting such a designe, so far as I can do it" (Weld, *op. cit.*, I, 306).

[141] Wolf, *op. cit.*, p. 468.

[142] Ball, *op. cit.*, p. 80.

[143] Yet Ball (*ibid.*, p. 74) says: "If we desired to find to whom the spread of the study of Newton is due, we must look among proctors, moderators, colleges, tutors who accepted his doctrine."

Whiston, a convert to his system. But at the same time, even until 1718, Cartesian physics was upheld by other professors. It is significant that the first special chair for a branch of natural philosophy, chemistry (which then comprised heat, electricity, and magnetism), was founded in 1702, outside the seventeenth century.[144] Even afterward in the eighteenth century, experimental science remained always subordinate to mathematics in Cambridge, and, as Ball puts it, there was destined to follow upon the reign of logic a reign of mathematics, no less absolute and one-sided.[145]

During the last decades of the century, traces of interest in science are to be found in most universities. The universities of St. Andrews and Edinburgh have the distinction of being the first places where Newton's ideas, through David Gregory, became the object of academic discussion.[146] The University of Glasgow in 1692 pointed to its need of apparatus and instruments, and in 1693 obtained a telescope with prisms and tubes from George Sinclair, professor of mathematics.[147] Edinburgh in 1695 established a special professorship of botany.[148] Thus it must be admitted that scientific inquiry in 1700 was beginning to assume a different aspect in the English colleges from that which it presented in 1600.

HOLLAND

The most promising conditions for university development during the seventeenth century existed undoubtedly in Holland. I therefore greatly regret my inability to obtain any account of their activities other than sporadic hints in books dealing with other subjects.

The first favorable element in the situation was that Dutch

---

[144] "Cambridge," *Encyclopedia of Education*.
[145] Ball, *op. cit.*, p. 253.
[146] Brewster, *op. cit.*, p. 265.   [147] Coutts, *op. cit.*, p. 195.
[148] A. Grant, *The story of the University of Edinburgh*, p. 217.

universities were for the most part city foundations and under city regulations.[149] Then the foundations of all these universities fall close to or within the seventeenth century, and hence they did not have the burden of a long inheritance of scholasticism to contend with. If, besides, we realize that at this period the Dutch cities were exceedingly wealthy, that in Amsterdam toleration, freedom of speech, freedom of the press, existed beyond anything conceived of in other parts of Europe; that there was the greatest appreciation of, and friendliness toward, scholarship, we can well understand why Dutch universities took a leading position among schools of learning. They were not so much Dutch as international institutions. Leyden was the university of the Huguenots and Puritans driven from their own countries, and was also much frequented by Germans.[150] Similarly, Utrecht and Groningen had many German and English students.[151]

Turning to those matters which are of special interest to us, we find, contrary to what would be expected, that there was no liberty of religious belief at these universities, and that at Leyden and Utrecht the Calvinistic doctrine was prescribed for students and professors.[152] The most progressive and most impor-

[149] The University of Leyden was founded in 1575 by William of Orange (Minerva, *op. cit.*, p. 164); the University of Groningen by the governing body of the province in 1614 (*ibid.*, p. 163). The Athenaeum Illustre, the University of Amsterdam, was founded by the city magistrates in 1634 (*ibid.*, p. 162); the University of Utrecht by the city through the efforts of its mayor in 1634 (*ibid.*, p. 165).

[150] Eulenburg (*op. cit.*, p. 127), calculates that at Leyden an average of five hundred German students studied yearly. Edward F. S. A. Peacock, in an *Index of English speaking students who have graduated at Leyden University*, collected the names of 4,300 English students who studied at Leyden up to 1835, but looking through the *Index* I found that they were mostly students of the seventeenth century.

[151] Groningen had, up to 1690, 3,548 Dutch and 2,683 foreigners, of whom 2,141 were Germans (Minerva, *op. cit.*, p. 164).

[152] Tholuck, *op. cit.*, p. 5.

tant faculties of these Dutch universities were those of law and medicine. The faculty of medicine at Leyden, under the guidance of the great iatrochemist, Francis Sylvius de la Boe (1614–72), showed that progress toward the cultivation of the scientific and practical study of medicine which seems lacking in other countries. He was a follower of Van Helmont's ideas, but built up his chemical studies on the much more scientific basis of a knowledge of anatomy and physiology. Realizing thus the vast significance of chemical research for medicine, he persuaded the curators of the university to build for him a "laboratorium, as they call it."[153] Besides, another fundamental innovation was introduced. To the University of Leyden the credit is due of having made clinical teaching a permanent institution, and of having through its students transmitted the custom to other places.[154] At Leyden the greatest students of medicine of

[153] Foster, *op. cit.*, p. 147.

[154] Puschmann (*op. cit.*, pp. 411 f.) gives the following account of the establishment of clinical teaching: "The Professors Otto Van Heurne and E. Schrevelius inaugurated this system about the year 1630 in the infirmary at Leyden. The method adopted was for the students first of all to examine the patient on his complaint, then for each one to state his view upon the nature, causes, symptoms, prognosis and treatment of the disease, and last of all for the Professor to confirm the correct opinion, to confute erroneous ones, and to add any explanation required. But this procedure did not please the students, for they ran the risk of having their ignorance exposed by questions which they could not answer, and O. V. Heurne found himself obliged reluctantly to give it up and in its place to undertake the examination of the patients himself, and to follow this up closely with directions for treatment. The bodies of the patients who died in the hospital were opened in order to arrive at certainty as to the cause and seat of their diseases. An apothecary's shop was also attached to this hospital where the students could see and learn how to prepare medicines.

"In 1648 Albert Kyper, to whom we owe this account, coming from Königsberg in Prussia, took over the direction of the clinic at Leyden. After a few years he was succeeded by F. de la Boe (Sylvius) who has been thus described, when engaged in clinical instruction, by his colleague Lucas Schacht; 'when he came with his pupils to the patient and began to teach, he appeared completely in the dark as to the causes or the nature of the affection the patient was suffering from, and at first expressed no opinion upon the case; he then began by questions put to different members of his audience to fish out [*expiscabatur*] every-

SCIENCE IN THE UNIVERSITIES

the century such as Swammerdam, Willis, and Steno[155] were educated; and it is worthy of note that Guericke, Hevelius, and Nehemiah Grew studied there, and that Huygens in his last will bequeathed his manuscripts to that university.

The progressive spirit of the Dutch universities as a whole is evinced by their attitude toward the Cartesian teaching. At Utrecht it was immediately taught by Renery, a pupil of Descartes, and by Henry Regius (Van Roy), professor of theoretical medicine. But even here it had to fight some battles with the orthodox theological party, voicing its sentiments through Voetius, the rector of the university, no less an opponent of Harvey's theory of circulation of blood than of Descartes.[156] At Leyden and Groningen it easily overthrew Aristotelian physics.

In the territory which constitutes Belgium today only one university flourished, namely, Louvain. About this, I unfortunately found out only that it was a city foundation of great prominence throughout the period, with a famous medical school, thronged by foreign students. To what extent, if any modern methods were introduced in the seventeenth century, I have not been able to ascertain.

Having thus touched upon some features of university development during the seventeenth century in the five leading countries of Europe, we are prepared to turn once more to the conditions of scientific progress enumerated at the opening of

---

thing, and finally united the facts discovered in this manner into a complete picture of the disease, in such a way that the students received the impression that they had themselves made the diagnosis and not learnt it from him.' Under his direction the Leyden clinic acquired such a reputation that 'students and doctors came thither,' as Schacht says, 'from Hungary, Russia, Poland, Germany, Denmark and Sweden, from Switzerland, Italy, France and England, in fact from every country in Europe.' . . . ."

[155] Foster, *op. cit.*, p. 174. Steno extols the great enthusiasm of Sylvius who impressed upon his students how little he knew, and how much remained to be discovered.

[156] Mullinger, *op. cit.*, III, 430.

this chapter, and ask ourselves how far the universities had begun to fulfil these conditions in the seventeenth century.

1. The statement that science can thrive only in universities where a secular spirit prevails is borne out by the fact that in the universities of North Italy and Holland, and in Altdorf—all controlled by municipalities—active centers of scientific progress were found; while in England, in France, and in most German universities, where in 1700 the ecclesiastical element was still pre-eminent, no such promising condition existed.

2. No change of methods of instruction had been introduced by 1700, but the medieval disputations were perpetuated, even in the University of Halle, founded at the end of the century.[157]

3. The emphasis on logic, metaphysics, and ancient languages was hardly lessened.

4. Sporadic instances occurred of the establishment of the various sciences as independent disciplines: in England, professorships of astronomy, botany, experimental physics; in Germany, a chair of chemistry and several of physics[158] and of botany—insignificant beginnings, but implying a tendency toward specialization.

5. In two instances, Leyden and Altdorf, universities had established laboratories from their own funds. Bologna obtained one through bequest. Laboratory work was occasionally carried on informally at the homes of professors; botanical gardens were generally established.

6. The new truths were very slowly incorporated into textbooks, as was seen in the case of the Newtonian philosophy.[159]

[157] Eulenburg, *op. cit.*, p. 225.

[158] Rosenberger, *op. cit.*, II, 77: "Even if brilliant physicists used occasionally the method of experiment, this had not brought it to general recognition and understanding. The chairs of physics in the universities were still occupied by peripatetic philosophers."

[159] Kopp (*op. cit.*, II, 182) says that no textbook of the time accepted Boyle's ideas but all clung to Paracelsus' views. Gerland-Traumüller (*op. cit.*, II, 201 ff.) state that textbooks on experimental physics were creations of the eighteenth century.

7. The reform of the study of medicine through the introduction of methods of dissection commenced at the end of the sixteenth century, was continued throughout the seventeenth, and brought to relatively high perfection at Padua, Leyden, and London. Anatomical theaters were established at many universities. Microscopic study was pursued, privately only, by a few professors. Clinical teaching existed only in Leyden. A polyclinic was established in connection with the University of Paris.

8. The faculties of theology and law retained their former position of pre-eminence.

9. Latin was still the official university language in all countries.[160] The vernacular was adopted at only one German university, Halle.

10. There is no evidence of university interest in technical studies.

11. The degree of freedom of thought in matters of philosophic and scientific inquiry can be tested by the attitude taken by the universities toward the Copernican and Cartesian ideas. Neither was permitted to be taught at any Catholic institution during the seventeenth century. Yet I find that even at the Protestant university of Tübingen, Michael Mästlin was afraid to teach the Copernican doctrine.[161] It would appear that not only at Catholic universities were certain beliefs officially banned, but that the teachings of the Bible established in Protestant institutions the furthest limit to which inquiry might be carried.

Freedom of conscience seems not to have existed except in the North Italian institutions.

12. Freedom of the press existed in 1700 in Holland and in England. In France the law of 1626 prescribing the death penalty to the author of a book that was not submitted to the censor was still in force in 1700; in Germany there was an imperial

---

[160] Paulsen, however, suggests that even if the teaching of Latin was the same in 1700 as in 1600, the belief in its indispensability had ceased.

[161] Reicke, *op. cit.*, p. 109.

censorship commission; in Catholic countries the Papal Congregation of the Index reinforced the activity of the civil censors.

It was not until the early years of the eighteenth century that the University of Halle adopted the principle of *libertas philosophandi*,[162] and that the regulations in Kiel declared (1707) "that no faculty be henceforth bound to certain principles and opinions as far as they depend on human authority, but that every teacher give a free and arbitrary examination of all truths."[163] Not until 1720 did the Faculty of Arts of Paris place Descartes' *Méthode* beside the *Organon* and the *Metaphysics* of Aristotle.

[162] Paulsen, *op. cit.*, I, 372.
[163] Eulenburg, *op. cit.*, p. 140.

CONCLUSION

The preceding review of university instruction in the seventeenth century leads to the conclusion that, with the exception of the medical faculties, universities contributed little to the advancement of science.

This is further substantiated by the fact that many of the greatest scientists and thinkers were without any affiliation with universities; in England, Bacon and Grew, Boyle, Flamsteed, Willoughby; in Holland, Huygens, Leeuwenhoek, Swammerdam, Van Helmont; in Germany, Hevelius, Leibniz, earlier Kepler, Guericke; in France, Descartes, Pascal, Mariotte, Lémery. Many others, while not entirely disassociated from the universities, were affiliated with them only during a brief and insignificant period of their life;[1] others again, as Robert Hooke and Gassendi, held professorships at institutions such as the Gresham College or the Collége de France, which cannot be counted among regular universities.

It is true that several of the greatest scientists occupied professorial chairs; but even in those instances it often can be shown that their efficiency and prominence were due to forces foreign to the universities.[2]

Nor was this separation of the majority of scientists from the university atmosphere accidental. It grew in most cases from a conviction that the type of work done there—indeed, the entire educational fabric—was valueless, or at least foreign to their aims. This conviction had arisen before the seventeenth

[1] See Papin.

[2] As was shown above, Newton had lectured to Cambridge students for three years on his novel theory of colors, and no mention of it had reached a body as eager for such information as the Royal Society. Halley occupied his professorship after he was well established in fame through a career disassociated from the universities.

century in the minds of Roger Bacon, Cardan, Campanella, Telesio, Montaigne, especially in Paracelsus,[3] and Peter Ramus;[4] but in the seventeenth century it became much more general and showed itself both in the open criticisms of many of the leading thinkers[5] and in the numerous schemes and projects of reform which fill the records of the educational histories of Germany and England of this period.

All criticisms against the universities were directed against the conservatism of the system. It is not our province here to examine how far such conservatism is inevitable and inherent in all established institutions. Only a few additional facts will be

[3] For an excellent account of Paracelsus' attitude see Meyer, *op. cit.*, IV, 425 ff.

[4] V. A. Huber, *The English Universities*, II, 292.

[5] In the previous pages the criticisms of Bacon, Descartes, Leibniz, Jungius, Hobbes, and Thomasius were noticed.

Contrast with the prevailing spirit of criticism of the universities the expression of admiration and loyalty by Sprat, the historian of the Royal Society (*op. cit.*, p. 328): "I confess there haven't been wanting some forward assertors of the new philosophy, who haven't used moderation towards them [universities] but have concluded that nothing can be well done in the new discoveries unless all ancient arts be first rejected and their nurseries abolished—" but, "the rashness of these men's proceedings has prejudiced rather than advanced what they make the show to promote—" . . . . "It is but just that we should have this tenderness for the interest of these magnificent seats of humane knowledge and divine, to which the natural philosophy of our nation cannot be injurious without horrible ingratitude, seeing in them it has been principally cherished and revised. It is true that such experimental studies are largely dispersed at this time. But they first came forth thence as colonies of old did from Rome."

Yet Sprat confesses (*ibid.*, p. 68): "Our seats of knowledge are not laboratories as they ought to be but only schools," and (*ibid.*, pp. 329 f.) "Nothing more suppresses the genius of learners than formality and confinement of precepts by which they are instructed. . . . . I venture to propose whether it were not as profitable to apply the eyes and hands of children to see and touch all the several sensible things as to oblige them to learn and remember the doctrines of general arts—whether mechanical education would not excel the methodical." . . . . "It is the memory which has most vigor in children and judgment in men . . . . [hence] we take a preposterous course in education by teaching general rules before particular things." Thus it would seem that Sprat's praise of the universities was rather that of loyalty than of approval.

CONCLUSION

adduced to show that this has been a salient characteristic of the great educational bodies. No change was made from 1570 to 1858 in the statutes of Oxford;[6] no essential change from 1558 to 1830 in the organization of the University of Leipzig;[7] and no change from 1360 to 1783 in the laws of the theological faculty of Bologna.[8] Indeed, the history of universities in the eighteenth century continues to be a record of conservatism. In Germany, for example, while, on the one hand, the University of Göttingen was founded in 1733 where science attained an ideal home, on the other hand, in 1740 the University of Innsbruck refused the establishment of a professorship in botany and chemistry and had the latter subject studied in apothecary shops;[9] and in Erlangen the professor of chemistry had to give all laboratory instructions in his own house and with his own apparatus, from 1754 to 1769.[10]

It thus would seem from the slight progress of the universities along lines of experimental science, from the fact that the greatest scientists of the age were not affiliated with them, from the many criticisms leveled against them, and from actual evidences of their conservatism, extending even into the eighteenth century, that the universities in the seventeenth century did not lend to science that encouragement which it needed in order to take root in them.

Let us turn from the university situation to that of the scientific societies. It is superfluous to say that they made every effort to foster the cause of experimental science. This was the keynote, the charter, of their existence, the motive underlying their every activity. Yet it may not be amiss to emphasize again the most characteristic directions of their efforts. These may be

[6] Minerva, *op. cit.*, I, 218.
[7] *Ibid.*, p. 30.
[8] Mazzetti, *op. cit.*, p. 46.
[9] Puschmann, *op. cit.*, p. 30.   [10] Günther, *op. cit.*, p. 6.

epitomized as follows: The societies concentrated groups of scientists at one place, performed experiments and investigations impossible to individual effort, encouraged individual scientists and gave them both opportunity and leisure, often through financial support, for scientific work. They became centers of scientific information (*fora sapientiae*), published and translated scientific books, promulgated periodically scientific discoveries, and thus co-ordinated the scientific efforts of the various progressive European countries. They concerned themselves about matters of homely interest such as trade, commerce, tools, and machinery, and tried to improve everyday life by the light of science. They contributed to the general enlightment by dispelling popular errors, and at times endeavored to reach the public by means of lectures. But first and foremost they developed the scientific laboratory, created the national observatory, devised, perfected, and standardized instruments, originated and insisted on exact methods of experimentation, and thus established permanently the laboratory method as the only true means of scientific study.[11]

[11] The great significance of scientific societies is by no means a consideration a posteriori but was felt by many scientists and thinkers during the seventeenth century. Gregory wrote to Collins (Rigaud, *op. cit.*, I, 158): "They [i.e., the Royal Society] will cause great changes throughout all the body of natural philosophy," and Weld (*op. cit.*, I, 422) relates that "Sir Isaac Newton was pleased to say he wished there were a Philosophical Society in every town where there was company to support them." Dr. E. Bernard wrote to Collins (Rigaud, *op. cit.*, I [April 3, 1671], 158 f.) referring to Oxford: "Books and experiments do well together, but separately they betray imperfection. . . . . The happy Royal Society adjusts both together, and I doubt not but in a short while will approve itself so great a friend and near ally to the universities that by the munificence of some of the members of noble fellowship there may be occasion given of frequent experiment in both famous universities and consequently of lasting commerce."

Boyle composed *A Treatise Written to Recommend the Whole Design of the Society* and honored the Society by leaving them in his will a collection . . . . "as a testimony of my great respect for the illustrious Society, ... wishing them . . . . a happy success in their laudable attempts to discover the true nature of the works of God" (Birch, *Works of the Honourable Robert Boyle,* p. clx).

CONCLUSION

The conclusion is thus inevitable that the organized support which science needed in order to penetrate into the thought and lives of people was not obtained from universities, but was derived from those forms of corporate activity which it had created for itself, the scientific societies.

Our interpretation of the relative contributions of universities and societies to the newly born sciences has been generally adopted by students of both, and it may be of interest in conclusion to quote some of their views. Harnack, in his study of the Berlin Academy, says:

> The European universities were born at the high tide of the Middle Ages, and their institutions corresponded to the medieval attitude of transmitting the sum of knowledge in fixed forms. The Academies of Europe belong to the epoch which begins in the middle of the seventeenth century, and their institutions are an expression of the new spirit which was thenceforth to attain its power in the realm of thought and life.[12]

Lexis, the historian of the German universities, says:

> The new philosophy and natural science was not born within the walls of the universities. The result of this fact was that the universities lagged behind the spirit of the time and gradually fell into disrepute.[13] . . . . While at the universities the old scholastic physics and cosmology was lectured upon in close adherence to the Aristotelian text, there arose outside and scorned by them, modern mathematical and scientific learning, which eventually led to a complete transformation of our conception of the universe. The universities remained unaffected . . . . except that they felt occasionally called upon to rise in defense against the invaders. Hence modern philosophy grew into a power hostile to the universities, which seemed superannuated in view of the newer educational ideals.[14]

Tannery, in his review of the progress of sciences at this period, says:

> The universities . . . . were incapable of transforming themselves to the needs of the new times. . . . . Hence the great necessity for societies

[12] Harnack, *op. cit.*, I, 5.
[13] Lexis, *op. cit.*, p. 4.
[14] *Ibid.*, p. 16.

capable of centralizing the efforts of scientific workers and of supplying funds for expensive experiments and for the expenses of scientific publications.[15]

Eulenburg, in his study of the German universities, writes:

The comparatively slight regard shown for the new studies by the universities leads to a higher estimation of academies. Great works and important discussions are to be found far oftener in the transactions of academicians than at the lectures of universities.[16]

To sum up: These societies were the *Kulturträger* of the second half of the seventeenth century much as the universities had been before the scientific revolution. They were the concentrated expression of the new spirit which was to gain the supremacy in the realm of thought and life. They typify this age drunk with the fulness of new knowledge, busy with the uprooting of superannuated superstitions, breaking loose from traditions of the past, embracing most extravagant hopes for the future. In their midst the spirit of minute scientific inquiry is developed; here the charlatanry and curiosity of the alchemist and magician are transformed into methodical investigation; here the critical faculty is developed so that the disclosure of an error is as important as the discovery of a new truth; here the minute fact is put as high—nay, higher—than generalization; here the individual scientist learned to be contented and proud to have added an infinitesimal part to the sum of knowledge; here, in short, the modern scientist was evolved.

The universities today have little more in common with those of the seventeenth century than the name, their general organization, and a few formalities, such as conferring degrees. The revolution in the universities which caused them to assimilate the changes sketched above, making of the university professor a modern scientist, has been the task of the two centuries which have elapsed since the seventeenth and in a most real

---

[15] * Tannery, *Histoire de France*, VI, 394.
[16] Eulenburg, *op. cit.*, p. 137.

sense is still the task of our own time. This revolution has made and is making universities homes of free thought, of scientific research and instruction, places where matters most intimately connected with everyday life are fostered.

It was the unmistakable and magnificent achievement of the scientific societies of the seventeenth century, not only to put modern science on a solid foundation, but in good time to revolutionize the ideals and methods of the universities and render them the friends and promoters of experimental science instead of the stubborn foes they had so long been.

APPENDIX

BACON'S "HOUSE OF SALOMON"

The End of our Foundation is the knowledge of Causes, and secret motions of things; and the enlarging of the bounds of Human Empire, to the effecting of all things possible.

The Preparations and Instruments are these. We have large and deep caves of several depths: the deepest are sunk six hundred fathom; and some of them are digged and made under great hills and mountains: so that if you reckon together the depth of the hill and the depth of the cave, they are (some of them) about three miles deep. For we find that the depth of a hill, the depth of a cave from the flat, is the same thing; both remote alike from the sun and heaven's beams, and from the open air. These caves we call the Lower Region. And we use them for all coagulations, indurations, refrigerations, and conservations of bodies. We use them likewise for the imitation of natural mines; and the producing also of new artificial metals, by compositions and materials which we use and lay there for many years. We use them also sometimes, (which may seem strange,) for curing of some diseases, and for prolongation of life in some hermits that choose to live there, well accommodated of all things necessary; and indeed live very long; by whom also we learn many things.

We have burials in several earths, where we put divers cements, as the Chineses do their porcellain. But we have them in greater variety, and some of them more fine. We have also great variety of composts, and soils, for the making of the earth fruitful.

We have high towers; the highest about half a mile in height; and some of them likewise set upon high mountains; so that the vantage of the hill with the tower is in the highest of them three miles at least. And these places we call the Upper Region: accounting the air between the high places and the low, as a Middle Region. We use these towers, according to their several heights and situations, for insolation, refrigeration, conservation; and for the view of divers meteors; as winds, rain, snow, hail, and some of the fiery meteors also. And upon them, in some places, are dwellings of hermits, whom we visit sometimes, and instruct what to observe.

We have great lakes both salt and fresh, whereof we have use for the fish and fowl. We use them also for burials of some natural bodies: for

APPENDIX

we find a difference in things buried in earth or in air below the earth, and things buried in water. We have also pools, of which some do strain fresh water out of salt; and others by art do turn fresh water into salt. We have also some rocks in the midst of the sea, and some bays upon the shore for some works wherein is required the air and vapour of the sea. We have likewise violent streams and cataracts, which serve us for many motions: and likewise engines for multiplying and enforcing of winds, to set also on going divers motions.

We have also a number of artificial wells and fountains, made in imitation of the natural sources and baths; as tincted upon vitriol, sulphur, steel, lead, brass, nitre, and other minerals. And again we have little wells for infusions of many things, where the waters take the virtue quicker and better than in vessels or basons. And amongst them we have a water which we call Water of Paradise, being, by that we do to it, made very sovereign for health, and prolongation of life.

We have also great and spacious houses, where we imitate and demonstrate meteors; as snow, hail, rain, some artificial rains of bodies and not of water, thunders, lightnings; also generations of bodies in air; as frogs, flies, and divers others.

We have also certain chambers, which we call Chambers of Health, where we qualify the air as we think good and proper for the cure of divers diseases, and preservation of health.

We have also fair and large baths, of several mixtures, for the cure of diseases, and the restoring of man's body from arefaction; and others for the confirming of it in strength of sinews, vital parts, and the very juice and substance of the body.

We have also large and various orchards and gardens, wherein we do not so much respect beauty, as variety of ground and soil, proper for divers trees and herbs; and some very spacious, where trees and berries are set whereof we make divers kinds of drinks, besides the vineyards. In these we practice likewise all conclusions of grafting and inoculating, as well of wild-trees as fruit-trees, which produceth many effects. And we make (by art) in the same orchards and gardens, trees and flowers to come earlier or later than their seasons; and to come up and bear more speedily than by their natural course they do. We make them also by art greater much than their nature; and their fruit greater and sweeter and of differing taste, smell, colour, and figure, from their nature. And many of them we so order, as they become of medicinal use.

We have also means to make divers plants rise by mixtures of earths without seeds; and likewise to make divers new plants, differing from the vulgar; and to make one tree or plant turn into another.

We have also parks and inclosures of all sorts of beasts and birds, which we use not only for view or rareness, but likewise for dissections and trials; that thereby we may take light what may be wrought upon the body of man. Wherein we find many strange effects; as continuing life in them, though divers parts, which you account vital, be perished and taken forth; resuscitating of some that seem dead in appearance; and the like. We try also all poisons and other medicines upon them, as well of chirurgery as physic. By art likewise, we make them greater or taller than their kind is; and contrariwise dwarf them, and stay their growth: we make them more fruitful and bearing than their kind is; and contrariwise barren and not generative. Also we make them differ in colour, shape, activity, many ways. We find means to make commixtures and copulations of different kinds; which have produced many new kinds, and them not barren, as the general opinion is. We make a number of kind, of serpents, worms, flies, fishes, of putrefaction; whereof some are advanced (in effect) to be perfect creatures, like beasts or birds; and have sexes, and do propagate. Neither do we this by chance, but we know beforehand of what matter and commixture what kind of those creatures will arise.

We have also particular pools, where we make trials upon fishes, as we have said before of beasts and birds.

We have also places for breed and generation of those kinds of worms and flies which are of special use; such as are with you your silk-worms and bees.

I will not hold you long with recounting of our brew-houses, bake-houses, and kitchens, where are made divers drinks, breads, and meats, rare and of special effects. Wines we have of grapes; and drinks of other juice of fruits, of grains, and of roots: and of mixtures with honey, sugar, manna, and fruits dried and decocted. Also of the tears or woundings of trees, and of the pulp of canes. And these drinks are of several ages, some to the age or last of forty years. We have drinks also brewed with several herbs, and roots, and spices; yea with several fleshes, and white meats; whereof some of the drinks are such, as they are in effect meat and drink both: so that divers, especially in age, do desire to live with them, with little or no meat or bread. And above all, we strive to have drinks of extreme thin parts, to insinuate into the body, and yet without all biting, sharpness, or fretting; insomuch as some of them put upon the back of your hand will, with a little stay, pass through to the palm, and yet taste wild to the mouth. We have

APPENDIX 267

also waters which we ripen in that fashion, as they become nourishing; so that they are indeed excellent drink; and many will use no other. Breads we have of several grains, roots, and kernels; yea and some of flesh and fish dried; with divers kinds of leavenings and seasonings: so that some do extremely move appetites; some do nourish so, as divers do live of them, without any other meat; who live very long. So for meats, we have some of them so beaten and made tender and mortified, yet without all corrupting, as a weak heat of the stomach will turn them into good chylus, as well as a strong heat would meat otherwise prepared. We have some meats also and breads and drinks, which taken by men enable them to fast long after; and some other, that used make the very flesh of men's bodies sensibly more hard and tough, and their strength far greater than otherwise it would be.

We have dispensatories, or shops of medicines. Wherein you may easily think, if we have such variety of plants and living creatures more than you have in Europe, (for we know what you have,) the simples, drugs, and ingredients of medicines, must likewise be in so much the greater variety. We have them likewise of divers ages, and long fermentations. And for their preparations, we have not only all manner of exquisite distillations and separations, and especially by gentle heats and percolations through divers strainers, yea and substances; but also exact forms of composition, whereby they incorporate almost, as they were natural simples.

We have also divers mechanical arts, which you have not; and stuffs made by them; as papers, linen, silks, tissues; dainty works of feathers of wonderful lustre; excellent dyes, and many others; and shops likewise, as well for such as are not brought into vulgar use amongst us as for those that are. For you must know that of the things before recited, many of them are grown into use throughout the kingdom; but yet if they did flow from our invention, we have of them also for patterns and principals.

We have also furnaces of great diversities, and that keep great diversity of heats; fierce and quick, strong and constant; soft and mild; blown, quiet; dry, moist; and the like. But above all, we have heats in imitation of the sun's and heavenly bodies' heats, that pass divers inequalities and (as it were) orbs, progresses, and returns, whereby we produce admirable effects. Besides, we have heats of dungs, and of bellies and maws of living creatures, and of their bloods and bodies; and of hays and herbs laid up moist; of lime unquenched; and such like. Instruments also which generate heat only by motion. And farther, places for strong insolations; and again, places under the earth, which by nature or art yield heat. These divers heats we use, as the nature of the operation which we intend requireth.

We have also perspective-houses, where we make demonstrations of all lights and radiations; and of all colors; and out of things uncolored and transparent, we can represent unto you all several colours; not in rainbows, as it is in gems and prisms, but of themselves single. We represent also all multiplications of light, which we carry to great distance, and make so sharp as to discern small points and lines; also all colorations of light: all delusions and deceits of the sight, in figures, magnitudes, motions, colours: all demonstrations of shadows. We find also divers means, yet unknown to you, of producing of light originally from divers bodies. We procure means of seeing objects afar off; as in the heaven and remote places; and represent things near as afar off, and things afar off as near; making feigned distances. We have also helps for the sight, far above spectacles and glasses in use. We have also glasses and means to see small and minute bodies perfectly and distinctly; as the shapes and colours of small flies and worms, grains and flaws in gems, which cannot otherwise be seen; observations in urine and blood, not otherwise to be seen. We make artificial rainbows, halos, and circles about light. We represent also all manner of reflexions, refractions, and multiplications of visual beams of objects.

We have also precious stones of all kinds, many of them of great beauty, and to you unknown; crystals likewise; and glasses of divers kinds; and amongst them some of metals vitrificated, and other materials beside those of which you make glass. Also a number of fossils, and imperfect minerals, which you have not. Likewise loadstones of prodigious virtue; and other rare stones, both natural and artificial.

We have also sound-houses, where we practise and demonstrate all sounds, and their generation. We have harmonies which you have not, of quarter-sounds, and lesser slides of sounds. Divers instruments of music likewise to you unknown, some sweeter than any you have; together with bells and rings that are dainty and sweet. We represent small sounds as great and deep; likewise great sounds extenuate and sharp; we make divers tremblings and warblings of sounds, which in their original are entire. We represent and imitate all articulate sounds and letters, and the voices and notes of beasts and birds. We have certain helps which set to the ear do further the hearing greatly. We have also divers strange and artificial echoes, reflecting the voice many times, and as it were tossing it: and some that give back the voice louder than it came; some shriller, and some deeper; yea, some rendering the voice differing in the letters or articulate sound from that they receive. We have also means to convey sounds in trunks and pipes, in strange lines and distances.

APPENDIX 269

We have also perfume-houses; wherewith we join also practices of taste. We multiply smells, which may seem strange. We imitate smells, making all smells to breathe out of other mixtures than those that give them. We make divers imitations of taste likewise, so that they will deceive any man's taste. And in this house we contain also a confiture-house; where we make all sweet-meats, dry and moist, and divers pleasant wines, milks, broths, and sallets, far in greater variety than you have.

We have also engine-houses, where are prepared engines and instruments for all sorts of motions. There we imitate and practise to make swifter motions than any you have, either out of your muskets or any engine that you have; and to make them and multiply them more easily, and with small force, by wheels and other means: and to make them stronger, and more violent than yours are; exceeding your greatest cannons and basilisks. We represent also ordnance and instruments of war, and engines of all kinds: and likewise new mixtures and compositions of gun-powder, wild-fires burning in water, and unquenchable. Also fire-works of all variety both for pleasure and use. We imitate also flights of birds; we have some degrees of flying in the air; we have ships and boats for going under water, and brooking of seas; also swimming-girdles and supporters. We have divers curious clocks, and other like motions of return, and some perpetual motions. We imitate also motions of living creatures, by images of men, beasts, birds, fishes, and serpents. We have also a great number of other various motions, strange for equality, fineness, and subtilty.

We have also a mathematical-house, where are represented all instruments, as well of geometry as astronomy, exquisitely made.

We have also houses of deceits of the senses; where we represent all manner of feats of juggling, false apparitions, impostures, and illusions; and their fallacies. And surely you will easily believe that we that have so many things truly natural which induce admiration, could in a world of particulars deceive the senses, if we would disguise those things and labour to make them seem more miraculous. But we do hate all impostures and lies; insomuch as we have severely forbidden it to all our fellows, under pain of ignominy and fines, that they do not shew any natural work or thing, adorned or swelling; but only pure as it is, and without all affectation of strangeness.

These are (my son) the riches of Salomon's House.

For the several employments and offices of our fellows; we have twelve that sail into foreign countries, under the names of other nations, (for our own we conceal;) who bring us the books, and abstracts, and patterns of experiments of all other parts. These we call Merchants of Light.

We have three that collect the experiments which are in all books. These we call Depredators.

We have three that collect the experiments of all mechanical arts; and also of liberal sciences; and also of practices which are not brought into arts. These we call Mystery-men.

We have three that try new experiments, such as themselves think good. These we call Pioners or Miners.

We have three that draw the experiments of the former four into titles and tables, to give the better light for the drawing of observations and axioms out of them. These we call Compilers.

We have three that bend themselves, looking into the experiments of their fellows, and cast about how to draw out of them things of use and practice for man's life, and knowledge as well for works as for plain demonstration of causes, means of natural divinations, and the easy and clear discovery of the virtues and parts of bodies. These we call Dowry-men or Benefactors.

Then after divers meetings and consults of our whole number, to consider of the former labours and collections, we have three that take care, out of them, to direct new experiments, of a higher light, more penetrating into nature than the former. These we call Lamps.

We have three others that do execute the experiments so directed, and report them. These we call Inoculators.

Lastly, we have three that raise the former discoveries by experiments into greater observations, axioms, and aphorisms. These we call Interpreters of Nature.

We have also, as you must think, novices and apprentices, that the succession of the former employed men do not fail; besides a great number of servants and attendants, men and women. And this we do also: we have consultations, which of the inventions and experiments which we have discovered shall be published, and which not: and take all an oath of secrecy, for the concealing of those which we think fit to keep secret: though some of those we do reveal sometimes to the state, and some not.

BIBLIOGRAPHY[1]

Academia Caesarea Leopoldina Carolina Germanica Naturae Curiosorum Miscellanea Curiosa sive Ephemeridum Medico Physicorum Germanorum (referred to as *Miscellanea Curiosa*).

Acta eruditorum. Lipsiae, 1682–1731.

Adam, Charles. *Philosophie de François Bacon.* Paris, 1890.

Agricola, George. *De re metallica libri XII.* Basileae, 1546.

Albini, Bernhardi. *Oratio de incrementis et statu artis medicae seculi decimi septimi.* Lugduni Batavorum, 1711.

Allbutt, Thomas C. *The historical relations of medicine and surgery to the end of the sixteenth century.* London, 1905.

Andreae, Johann Valentin. *Reipublicae Christianopolitanae descripto.* Argentorati, 1642.

Antinori, Vincenzio. *Notizie istoriche relative all'Accademia del Cimento.* Florence, 1841. (In *Accademia del Cimento, Saggi di naturali esperienze*, 1841.)

Arnoldt, D. H. *Historie der Königsbergischen Universität.* 2d ed. Königsberg, 1746.

Atti e Memorie. See Tozzetti.

Aubrey, John. *'Brief lives,' chiefly of contemporaries, set down by John Aubrey, between the years 1669 and 1696; edited from the author's MSS.* 2 vols. Oxford, 1898.

Bacon, Francis. *The philosophical works of Francis Bacon, baron of Verulam, Viscount of St. Albans . . . . reprinted from the texts and translations, with notes and prefaces, of Ellis and Spedding, edited with an introduction by John M. Robertson.* London, 1905.

von Baer, Karl E. *Reden gehalten in wissenschaftlichen Versammlungen.* 3 vols. 2d ed. Braunschweig, 1886.

Ball, W. W. R. *A short account of the history of mathematics.* Cambridge, 1889.

Barthold, Friedrich W. *Geschichte der fruchtbringenden Gesellschaft.* Berlin, 1848.

Bartholini, Thomae. *Acta medica et philosophica Hafniensia.* 1671–1769.

---

[1] Wherever before a reference an asterisk is found it has been impossible for the Editor to verify this reference in the libraries which were available.

Bartholomaei, Dr. F. "Erhard Weigel. Ein Beitrag zur Geschichte der mathematischer Wissenschaften auf den deutschen Universitäten im 17. Jahrhundert" *Zeit. für Mathematik und Physik* (1868), Vol. XIII. (Supplement.)

Bayle, Pierre. *Nouvelles de la republique des lettres; (Lettres à sa famille; Apologie de M. Bayle ou lettre d'un sceptique)*. Vol. I of *Œuvres diverses excepte son dictionnaire historique et critique*. Nouv. éd. 4 vols. La Haye, 1737.

Bertrand, Joseph L. F. *L'Académie des Sciences et les Académiciens de 1666 à 1793*. Paris, 1869.

Birch, Thomas. *History of the Royal Society of London*. 4 vols. London, 1756.

Birch, Thomas. *The Works of the Honourable Robert Boyle in five volumes. To which is prefixed the Life of the Author*. London, 1744.

Bock, Otto. *Die Reform der Erfurter Universität während des dreissigjährigen krieges*. Halle, 1908.

Bonet, Theophilus. *Medicina septentrionalis collatitia, sive rei medicae, nuperis annis a Medicis Anglis, Germanis et Danis emissae sylloge et syntaxis*. Geneva, 1686–1687.

Bosscha, Johannes. *Christian Huygens; Rede am 2oosten Gedächtnistage seines Lebensendes . . . . aus dem Holländischen übersetzt von T. W. Engelmann*. Leipzig, 1895.

Boyle, Robert. See Birch, Thomas.

Brewster, David. *The Life of Sir Isaac Newton*. London, 1831.

Büchner, A. E. *Academiae Sacri Romani Imperii Leopoldino-Carolinae. Naturae Curiosorum historia*. Halle, 1755.

*Büchner, A. E. *Sacrae Caesareae Majestatis Mandato et Privilegio Leges*. Halle, 1756.

Caesalpinus, Andreas. *De Plantis libri XVI*. Florence, 1583.

Cajori, Florian. *A history of physics in its elementary branches, including the evolution of physical laboratories*. New York and London, 1899.

de Candolle, Alphonse L. P. P. *Histoire des sciences et des savants depuis deux siècles suivie d'autres études sur des sujets scientifiques en particulier sur la sélection dans l'éspèce humaine*. Genève, 1873.

Cantor, Moritz. *Vorlesungen über Geschichte der Mathematik*. 4 vols. Leipzig, 1880–1908.

Carus, Julius V. *Geschichte der Zoologie bis auf Joh. Müller und Charl. Darwin*. München, 1872.

Carutti, Domenico. *Breve storia della Accademia dei Lincei*. Rome, 1883.

Cassini, Giovanni D. "De l'origine et du progres de l'astronomie et de son usage dans la geographie et dans la navigation," *Memoires de l'Académie Royale des Sciences contenants les Ouvrages adopté es par cette Académie avant son renouvellement en 1699.* Amsterdam, 1736.

Chamberlayne, John. *The lives of the French, Italian and German philosophers, late members of the Royal Academy of Sciences in Paris. Together with abstracts of some of the choicest pieces, communicated by them to that illustrious society. To which is added the preface of the ingenious Monsieur Fontonelle, secretary, and author of the history of the said academy.* London, 1717.

Clark, Andrew. *The life and times of Anthony Wood, antiquary of Oxford, 1632–1695, described by himself; collected from his diaries and other papers.* 5 vols. Oxford, 1891–1900.

Collegium Curiosum sive Experimentale. Altdorf. 2 vols. of *Proceedings.* 1676, 1685.

Collins, John. *Commercium epistolicum.* 1712. (Published by the Royal Society.)

*Compayré, Gabriel. *Histoire critique des doctrines de l'éducation en France depuis le seizième siècle.* 4th ed. Paris, 1883.

Condorcet, M. J. *Œuvres complétes de Condorcet.* 21 vols. Brunswick, 1804. Vol. I: *Éloges des Académiciens de l'Académie Royale des Sciences morts depuis l'an 1666, jusqu'en 1699.*

Coutts, James. *A history of the University of Glasgow, from its foundation in 1451 to 1909.* Glasgow, 1909.

Daniel, C. H., and Barker, W. R. *Oxford University College Histories. Worcester College.* London, 1900.

Dannemann, F. *Grundiss einer Geschichte der Naturwissenschaften.* 2 vols. Leipzig, 1896–98.

Dannemann, F. *Die Naturwissenschaften in ihrer Entwicklung und in ihrem Zusammenhange*, Vol. II. Leipzig, 1911.

Daremberg, C. V. *Histoire des sciences médicales comprenant l'anatomie, la physiologie, la médicine, la chirurgie et les doctrines de pathologie générale.* Paris, 1870.

Della Porta, Joh. Baptista. *Magiae Naturalis libri XX.* Neapoli, 1589.

Dernburg, Heinrich. *Thomasius und die Stiftung der Universität Halle.* . . . . Halle, 1865.

Descartes, René. *The Philosophical works of.* 2 vols. (Translated by E. S. Haldane and G. R. T. Ross.) Cambridge, 1911.

Dictionary of National Biography.

Dircks, Henry. *The life, times and scientific labours of the second Mar-

quis of Worcester. To which is added a reprint of his Century of Inventions, 1663, with a commentary thereon. London, 1865.

Duhamel, J. B. *Regiae scientiarum academiae historia.* . . . . Leipzig, 1700.

Encyclopaedia Britannica. 11th ed.

Encyclopedia of Education.

Eulenburg, Franz. "Die Frequenz der deutschen Universitäten von ihrer Gründung bis zur Gegenwart," *Abhandlungen der Königl. Sächs. Gesellschaft der Wissenschaften* ("Philologisch-Historische Klasse," Vol. XXIV[1]), Vol. LIII[2]. Leipzig, 1904.

Fernow, H. *Das Royal College of Physicians in London. Ein Beitrag zur englischen Kulturgeschichte.* Hamburg, 1905.

Fink, K. A. *A brief history of mathematics. An authorized translation by W. W. Beman and D. E. Smith.* Chicago, 1900.

*Fischer, Kuno. *Gottfried Wilhelm Leibniz, leben, werke und lehre.* 3rd ed. Heidelberg, 1889.

Fischer, Kuno. *Francis Bacon und seine Schule, Entwicklungs-geschichte der Erfahrungs-philosophie.* 3rd ed. Heidelberg, 1904.

Fontenelle, B. *Œuvres de Fontenelle, précédées d'une notice sur sa vie et ses ouvrages.* Paris, 1825. 5 vols. Vol. I–II: *Éloges.*

Foster, Sir Michael. *Lectures on the history of physiology during the sixteenth, seventeenth and eighteenth centuries.* Cambridge, 1901.

Foucher de Careil, A. *Œuvres de Leibniz publiées pour la première fois d'après les manuscrits originaux avec notes et introductions,* Vol. VII: *Leibniz et les académies.* Paris, 1875.

Fowler, Thomas. *Bacon's Novum organum.* 2d ed. Oxford. 1889.

Franqueville, A. C. F. *Le premiere siècle de l'Institut de France (25 octobre, 1795–25 octobre, 1895).* 2 vols. Paris, 1895–96.

Freheri, Pauli. *Theatrum virorum eruditione clarorum.* Nuremberg, 1688.

Galilei, Galileo. *Discorsi e demonstrazioni matematiche intorno a due nuove scienze.* (Ostwald's Klassiker der exakten Wissenschaften, Nos. 11, 24, and 25. 1890–1891. 'Unterredungen und mathematische Demonstrationen über zwei neue Wissenszweige, die Mechanik und die Fallgesetze betreffend von Galileo Galilei.') Leipzig.

*Galilei, Galileo. *Siderus nuntius magna longeque admirabilia spectacula pandens.* London, 1653.

Galilei, Galileo. *Dialogo . . . . sopra i due massimi sistemi del mondo Tolemaico e Copernicano.* Florence, 1632.

Gassendi, Pierre. *The Mirrour of true Nobility and Gentility. Being the Life of the Renowned Nicolaus Claudius Fabricius, Lord of Pieresk,*

Senator of the Parliament at Aix. (Englished by W. Rand.) London, 1657.
Geiger, Georg. *Geographische Studien an der Universität Altdorf.* Borna-Leipzig, 1908.
Gerland, E., und Traumüller, F. *Geschichte der physikalischen Experimentier-kunst.* Leipzig, 1899.
Giessen. *Die Universität Giessen von 1607 bis 1907; Beiträge zu ihrer Geschichte.* 2 vols. Giessen, 1907.
Gilbert, William. *De magnete, magneticisque corporibus, et de magno magnete tellure; physiologie nova plurimis.* London, 1600.
Gilbert, William. *De mundo nostro sublunari philosophia nova.* Amsterdam, 1651.
Giornale de' letterati d'Italia. 40 vols. Venice, 1710–40.
Glanvill, Joseph. *Plus ultra; or, the Progress and Advancement of Knowledge since the days of Aristotle. In an Account of some of the most remarkable late improvements of Practical, Useful Learning: To Encourage Philosophical Endeavours. Occasioned by a Conference with one of the Notional Way.* London, 1668.
Glanvill, Joseph. *A Praefatory Answer to Mr. Henry Stubbe, The Doctor of Warwick, wherein the Malignity, Hypocrisie, Falsehood of his Temper, Pretenses, Reports, and the Impertinency of his Arguings and Quotations in his Animadversions on Plus ultra, are discovered.* London, 1671.
Glanvill, Joseph. *Scepsis scientifica: or, Confest ignorance, the way to science; in an essay of the vanity of dogmatizing, and confident opinion. With a reply to the exceptions of the learned Thomas Albius.* London, 1665.
Grant, A. *The story of the University of Edinburgh during its first three hundred years.* 2 vols. London, 1884.
Grew, Nehemiah. *Musaeum Regalis Societatis; or, a catalogue and description of the natural and artificial rarities belonging to the Royal Society and preserved at Gresham College . . . . Comparative anatomy of stomach and guts.* London, 1681.
Gudger, Eugene W. "George Marcgrave, the first student of American natural history," *Popular Science Monthly,* LXXI (1912), 250.
Guericke, Otto. *Experimenta nova (ut vocantur) Magdeburgica de vacuo spatio.* Amsterdam, 1672. (Ostwald's Klassiker der exakten Wissenschaften, No. 59. Leipzig, 1894. "Neue 'Magdeburgische' Versuche über den leeren Raum, 1672.")

Guhrauer, Gottschalk E. *Gottfried Wilhelm Freiherr v. Leibnitz. Eine Biographie . . . . mit neuen Beilagen und einem Register.* 2 vols. Breslau, 1846.
Guhrauer, Gottschalk E. *Joachim Jungius und sein Zeitalter . . . . nebst Goethe's fragmenten über Jungius.* Stuttgart, 1850.
Gundling, N. H. *Historie der Gelahrtheit, oder ausführliche Discourse . . . . über seine eigenen Positiones als auch . . . . über D. C. A. Heumanni Conspectum Republicae Literariae gehalten.* Frankfurt und Leipzig, 1734.
Günther, S. "Die mathematischen und Naturwissenschaften an der nürnbergischen Universität Altdorf," *Mitteilungen des Vereins für Geschichte der Stadt Nürnberg.* 3 vols. Nuremberg, 1881.
Haeser, H. *Lehrbuch der Geschichte der Medizin und der epidemischen Krankheiten.* 3 vols. Jena, 1875–82.
Hall, John. *An Humble Motion to the Parliament of England concerning the Advancement of Learning and Reformation of the Universities.* London, 1649.
Harnack, Adolf. *Geschichte der Königlich Preussichen Academie der Wissenschaften zu Berlin.* 3 vols. Berlin, 1900.
Harvey, William. *On the motion of the heart and blood in animals.* (Willis' translation, revised and edited by Alex. Bowie.) London, 1889.
Hatin, Louis E. *Histoire politique et littéraire de la presse en France, avec une introduction historique sur les origines du journal, et la bibliographie générale des journaux depuis leur origine.* 7 vols. Paris, 1859–61.
Hautz, Johann F. *Geschichte der Universität Heidelberg.* Mannheim, 1862–64.
Henrich, Ferd. "Über alte chemische Geräte, Öfen und Arbeitsmethoden," *Chemiker Zeitung*, XXXV (1911), 197.
Henrich, Ferd. "Aus Erlangens chemischer Vergangenheit," *Sitzungsberichte der physikalischen medizinischen Societät zu Erlangen*, XXXVIII (1906), 103.
Histoire de l'Académie des Sciences depuis son éstablissement en 1666 jusqu'à 1699. 11 vols. Paris, 1733.
Høffding, Harald. *Geschichte der neuren Philosophie. . . . .* Leipzig, 1895.
*Hogg, Jabez. *The Microscope: its History, Construction, and Application, being a familiar introduction to the use of the instrument and the study of microscopical science.* London, 1883.

Hooke, Robert. *An Attempt to Prove the Motion of the Earth by Observations*. London, 1674.

Hooke, Robert. *Micrographia: or some Physiological Descriptions of Minute Bodies made by Magnifying Glasses with Observations and Inquiries thereupon*. London, 1665.

Hooke, Robert. *Observations of the late Eminent Dr. Robert Hooke, S.R.S. and Geom. Prof. Gresham, and other Eminent Virtuoso's of his Time*. Published by W. Derham, F.R.S. London, 1726.

Huber, V. A. *The English Universities*. (From the German of V. A. Huber—an abridged translation, edited by Francis W. Newman.) London, 1843.

Huygenii, Christiani. *Opera varia*. Vol. I. Lugduni Batavorum, 1724.

Huxley, Thomas. *On the advisableness of improving natural knowledge*. 1866.

Jöcher, Christian G. *Allegemeines Gelehrten Lexicon*. 4 vols. Leipzig, 1750–51.

Jourdain, Charles M. *Histoire de l'Université de Paris au 17e et au 18e siècle*. Paris, 1862–66.

Journal des Sçavans. 1665–1705.

Kaufmann, Georg H. *Die Geschichte der deutschen Universitäten*. 2 vols. 1888–96.

Keller, Ludwig. *Die deutschen Gesellschaften des 18. Jahrhunderts und die moralischen Wochenschriften. Ein Beitrag zur Geschichte des deutschen Bildungslebens*. Berlin, 1900. (Vorträge und Aufsätze der Comenius-gesellschaft. 8. Jahrg. 2 Stück.)

Keller, Ludwig. *Comenius und die Akademien der Naturphilosophen des 17. Jahrhunderts*. Berlin, 1895. (Vorträge und Aufsätze aus der Comenius-gesellschaft. 3. Jahrg. 1. Stück.)

Keller, Ludwig. *Gottfried Wilhelm Leibniz und die deutschen Societäten des 17. Jahrhunderts*. Berlin, 1903. (Vortrage und Aufsätze aus der Comenius-gesellschaft. 11. Jahrg. 3 Stück.)

Kepler, Johann. *Dioptrice, seu demonstratio eorum quae visui et visibilibus propter Conspicilla non ita pridem inventa accidunt*. . . . . Augustae Vindelicorum, 1611.

Kircher, Athanasius. *Mundus subterraneus in XII libros digestus*. . . . . 2 vols. Amsterdam, 1665.

Klee, G. F. *Die Geschichte der Physik an der Universität Altdorf bis zum Jahr 1650*. Erlangen, 1908.

*Klopp, Onno. *Die Werke von Leibniz*. 1864–73.

Klüpfel, K. A. *Geschichte und Beschreibung der Universität Tübingen.* Tübingen, 1849.

Kobell, F. *Geschichte der Mineralogie, von 1650–1860.* Munich, 1864.

Kopp, Hermann. *Die Alchemie in älterer und neuerer Zeit; ein Beitrag zur Culturgeschichte.* Heidelberg, 1886.

Kopp, Hermann. *Über den Zustand der Naturwissenschaften in dem Mittelalkter.* Heidelberg, 1869.

Kopp, Hermann. *Beiträge zur Geschichte der Chemie.* 3 parts. Braunschweig, 1869.

Kopp, Hermann. *Geschichte der Chemie.* 4 vols. Braunschweig, 1843–47.

La Grande Encyclopédie.

Leibniz. See Foucher de Careil.

Leixner, Otto. "Aus der Vergangenheit des deutschen Zeitschriftenwesens. I, II, III," *Deutsche Revue*, Vol. VI (1881); Part I, Heft 5, p. 247; Part II, Heft 8, p. 250; Part III, Heft 9, p. 392.

Lexis, W. *Das Unterrichtswesen im Deutschen Reich.* Vol. I: *Die Universitäten im Deutschen Reich;* Vol. III: *Das Volksschulwesen und das Lehrerbildungswesen im Deutschen Reich;* Vol. IV, Part I: *Die technischen Hochschulen im Deutschen Reich.* Berlin, 1904.

Machines et inventions approuvées par l'Académie royale des sciences, depuis son établissement jusqu'à présent; avec leur déscription. Dessinées et publiées du consentement de l'Académie par M. Gallon. 1666/1701–1724/54. 7 vols. Paris, 1735–(77).

Mädler, J. H. *Geschichte der Himmelskunde von der ältesten bis auf die neueste Zeit.* Braunschweig, 1873.

Maindron, M. E. "Histoire des Sciences—L'Academie des Sciences. Sa fondation, ses anciens réglements, ses installations successive, ses collections, bibliographie de l'Academie," *La Revue Scientifique de la France et de l'etranger.* 1881. 3d series. Vol. I. Pp. 684 and 705.

Maury, Louis F. A. *Les académies d'autrefois. L'áncienne Académie des sciences.* Paris, 1864.

Mazzetti, Serafino. *Memorie storiche sopra l'Università e l'Instituto delle Scienze di Bologna, e sopra gli stabilimenti ed i corpi scientifici alla medesima addetti.* Bologna, 1840.

Mazzetti, Serafino. *Repertorio de tutti i Professori antichi e moderni della famosa Università e del celebre Instituto delle Scienze di Bologna.* Bologna, 1847.

Mencke, Johann B. *De charlataneria eruditorum declamationes duae, cum notis variorum.* . . . . Amsterdam, 1716.

Mersenne, M. *La verité des sciences.* T. du Bray, 1625.

BIBLIOGRAPHY

Mersenne, M. *Les questions théologiques, physiques, morales et mathematiques, où chacun trouvera du contentement, ou de l'exercise.* Paris, 1634.

Meyer, E. H. F. *Geschichte der Botanik.* 4 vols. Königsberg, 1854–57.

Minerva. *Handbuch der gelehrten Welt. Die Universitäten und Hochschulen—ihre Geschichte und Organization.* Strassburg, 1911.

Miscellanea Berolinensia ad incrementum scientarum. 7 vols. Berlin and Halle, 1710–37.

Miscellanea Curiosa. See *Academia Caesarea Leopoldina Carolina,* etc. Frankfurt and Leipzig 1670–1706; Nuremberg, 1695–1713.

Mullinger, James B. *The University of Cambridge,* Vol. III. Cambridge, 1911.

Museum, Amtlicher. *Führer durch die Sammlungen.* . . . . Munich, 1925.

*Museum, . *Die Kunst- und Kulturgeschichtlichen Sammlungen des germanischen.* . . . . Nuremberg, 1912.

van Musschenbroek, P. *Tentamina experimentorum naturalium captorum in Accademia del Cimento.* Lugdini Batavorum, 1731.

Neigebaur, J. D. F. *Geschichte der Kais. Leopoldino-Carolinischen Akademie der Naturforscher (während des zweiten Jahrhunderts ihres Bestehens).* Jena, 1860.

Neudoerffer, J. *Nachrichten von den vornehmsten Künstlern und Werkleuten so innerhalb hundert Jahren in Nürnberg gelebt haben 1546, nebst der Fortsetzung von A. Gulden 1660.* Nürnberg, 1828.

Olivet, Abbé. See Thoulier, P. J.

Oxford, Register of the University of, Vol. II (1571–1622). (Edited by Andrew Clark.) Oxford, 1887.

Pagel, Julius L. *Geschichte der Medicin.* 2 vols. Berlin, 1898.

Pagel, Julius L. "Historisch-Medicinische Bibliographie für die Jahre 1875–1896," *Geschichte der Medicin.* 2 vols. Berlin, 1898.

Pagel, Julius L. *Zeittafeln zur Geschichte der Medizin.* Berlin, 1908.

Papin, Denis. *D. Papin, sa vie et son œuvre.* By Baron A. A. Ernouf. Paris, 1874.

Pascal, Blaise. *Récit de la grande expérience de l'équilibre des liqueurs.* Paris, 1648.

*Paulsen, Friedrich. *Geschichte des gelehrten Unterrichts auf deutschen Universitäten.* Braunschweig, 1906.

Peacock, Edward. *Index to English speaking students who have graduated at Leyden University.* London, 1883.

Pelisson, Fontanier P. *Histoire de l'Académie Française jusqu'en 1652.* Paris, 1653.

Peters, Hermann. *Der Arzt und die Heilkunst in der deutschen Vergangenheit.* Leipzig, 1900.

Petri, R. J. *Das Mikroskop, von seinen Anfängen bis zur jetzigen Vervollkommung.* Berlin, 1896.

Petty, Sir William. *Advice of William Petty to Mr. Samuel Hartlib, for the advancement of some particular parts of learning.* London, 1647.

The Philosophical Transactions of the Royal Society of London, from their commencement, in 1665 to the year 1800. (Abridged, with notes and biographic illustrations by Charles Hutton, George Shaw, Richard Pearson.) Vol. I. 1665 to 1672. London, 1809.

Pinkerton, John. *A General Collection of the best and most interesting voyages and travels in all parts of the world, many of which are now translated into English.* 17 vols. 1808–14.

Poggendorf, J. C. *Geschichte der Physik.* Leipzig, 1879.

Poggendorf, J. C. *Lebenslinien zur Geschichte der exacten Wissenschaften seit Wiederherstellung derselben.* Berlin, 1853.

Poggendorf, J. C. *Biographisch-literarisches Handwörtenbuch zur Geschichte der exacten Wissenschaften.* Leipzig, 1863–1904.

Poppe, A. *Alphabetisch chronologische Uebersicht der Erfindungen und Entdeckungen auf dem Gebiete der Physik, Chemie, ...* 3d ed. Frankfurt, 1881.

Prutz, R. E. *Geschichte des deutschen Journalismus.* Hanover, 1845.

Puschmann, Theodor. *A history of medical education from the most remote to the most recent times.* (Translated and edited by Evan H. Hare.) London, 1891.

Putnam, G. Haven. *The censorship of the Church of Rome and its influence upon the production and distribution of literature. A study of the history of the prohibitory and expurgatory indexes, together with some consideration of the effects of Protestant censorship and of censorship by the state.* 2 vols. New York. 1906–7.

"Recueil d'observations faites en plusieurs voyages par ordre de Sa Majesté, pour perfectionner l'astronomie et la geographie. Avec divers traitez astronomiques. Par messieurs de l'Academie Royale des Sciences." Paris, 1693.

Reicke, Emil. *Der Gelehrte in der deutschen Vergangenheit.* Leipzig, 1900.

Rigaud, S. J. *Correspondence of scientific men of the seventeenth century, including letters of Barrow, Flamsted, Wallis and Newton, printed from the originals in the collection of . . . . . the earl of Macclesfield.* 2 vols. Oxford, 1841.

Robertson, D. Maclaren. *A history of the French Academy (1635[4]–*

1910). With an outline sketch showing its relation to its constituent academies. New York, 1910.

Rosenberger, F. *Die Geschichte der Physik in Grundzügen mit synchronistischen Tabellen der Mathematik, der Chemie und beschreibenden Naturwissenschaften sowie der allgemeinen Geschichte.* Braunschweig, 1882–90.

Royal Society of London, Record of the, Nos. 1 and 2. 1897–1901. (No. 2 is called 2d ed.)

Sabatier, J. C. *Researches historiques sur la Faculté de Médicine de Paris, depuis son origine jusqu'a nos jours.* Paris, 1837.

Sachs, J. *History of botany (1530–1860)* (Authorized translation by H. E. F. Garnsey; revised by Isaac B. Balfour. Oxford, 1890.

Salomon, Ludwig. *Geschichte des deutschen Zeitungswesens von den ersten Anfängen bis zur Wiederaufrichtung des deutschen Reiches.* 3 vols. Oldenburg, 1900–1906.

Schelenz, Hermann. *Zur Geschichte der pharmazeutisch-chemischen Destilliergeräte.* Berlin, 1911.

Schott, Gaspard. *Magia universalis naturae et artis sive recondita naturalium et artificialium rerum scientia.* Bambergae, 1677.

Schott, Gaspard. *Mechanica hydraulico-pneumatica.* Würzburg, 1657.

Schreiber, Heinrich. *Geschichte der Albert-Ludwigs-Universität zu Frieburg.* Freiburg. 3 vols. 1857–68.

Schwenter, Daniel. *Deliciae physico-mathematicae oder mathematicae und philosophische Erquickstunden.* Nuremberg, 1651–53.

Servus, Herman. *Die Geschichte des Fernrohrs bis auf die neueste Zeit.* Berlin, 1886.

Sprat, Thomas. *The History of the Royal Society of London, for the Improving of Natural Knowledge.* London, 1667. (2d ed., 1702.)

Strauss, Emil. *Dialog über die beiden hauptsächlichsten Weltsysteme, das Ptolemäische und das Kopernikanische von Galileo Galilei. Aus dem Italienischen übersetzt und erläutert, von Emil Strauss.* Leipzig, 1891.

Stubbs, J. W. *The history of the University of Dublin from its foundation to the end of the eighteenth century; with an appendix of original documents which, for the most part, are preserved in the college.* Dublin and London, 1889.

*Swainson, William. *Taxidermy with Biography of Zoölogists. Cabinet Cyclopedia conducted by Dionysius Lardner,* Vol. CXXIV(?). *Popular Science Monthly,* LXXXI, p. 252.

Tholuck, August. *Das akademische Leben des siebzehnten Jahrhunderts, mit besonderer Beziehung auf die protestantisch theologischen Fakultäten Deutschlands.* . . . . Halle, 1854.

Thomasius, Christian. *Von Nachahmung der Franzosen.* ("Deutsche Litteraturdenkmale des 18. und 19. Jahrhunderts," No. 51, neue folge, No. 1.) Stuttgart, 1894.

Thomasius, Christian. *Monatgespräche. Scherz- und Ernsthafter, Vernünftiger und Einfältiger Gedanken, über allerhand lustige und nützliche Bücker und Fragen.* Frankfurt and Leipzig, 1688.

Thomson, Thomas. *History of the Royal Society from its institution to the end of the eighteenth century.* London, 1812.

Thoulier, Pierre Joseph, originally P. J. Olivet, known as Abbé Olivet. *Histoire de l'Académie Française, depuis 1652 juspu'à 1700.* Paris, 1730.

Tiraboschi, Girolamo. *Storia della letteratura Italiana.* 9 vols. Roma, 1782-85.

Tozzetti, G. Targioni. *Atti e Memorie inedite dell'Accademia del Cimento.* . . . . 3 vols. Firenze, 1780.

Traill, Henry D. *Social England; a record of the progress of the people in religion, laws, learning, arts, industry, commerce, science, literature and manners, from the earliest times to the present day,* Vol. V. London, 1900.

Treitschke, R. *B. Mencke, Professor der Geschichte zu Leipzig.* Leipzig, 1842.

Vesalius, Andreas. *De Humani Corporis Fabrica.* Basileae, 1543.

Vindiciae Academiarum; containing some briefe animadversions upon Mr. Webster's book, stiled, The Examination of Academies. Together with an Appendix concerning what M. Hobbes and M. Dell have published on this Argument. London, 1654.

Vockerodt, Gottfried. *Excercitationes academicae; sive, Commentatio eruditorum Societatibus et varia re litteraria, nec non philologemata sacra, auctius et emendatius edita.* Gotha, 1704.

Voltaire, François M. A. *Letters concerning the English Nation.* London, 1733.

Waller, Richard. *Essayes of natural experiments, made in the Academie del Cimento, under the protection of the most serene Prince Leopold of Tuscany.* (Written in Italian by the secretary of that academy. Englished by Richard Waller.) London, 1684.

Wallis, John. *Arithmetica Infinitorum.* Oxford, 1655.

Wallis, John. *Commercium Epistolicum, de Quaestionibus quibusdam Mathematicis nuper habitum.* Oxford, 1658.

Warburton, B. E. G. (known as Elliot Warburton). *Memoirs of Prince*

BIBLIOGRAPHY

Rupert and the Cavaliers, with their Private Correspondence. 3 vols. London, 1849.

Webster, J. *Academiarum Examen; or, the Examination of Academies; Wherein is discussed and examined the matter, method, and customes of academick and scholastick learning, and the insufficiency thereof discovered and laid open; as also some expedients proposed for the reforming of schools, etc.* London, 1654.

Wegele, Franz X. *Geschichte der deutschen Historiographie seit dem Auftreten des Humanismus.* Munich, 1885.

Wegele, Franz X. *Geschichte der Universität Würzburg.* 2 vols. Würzburg, 1882.

Weld, C. R. *A history of the Royal Society with Memoirs of the Presidents, compiled from Authentic Documents.* London, 1848.

Wells, Joseph. *Oxford University College Histories. Wadham College.* London, 1898.

Whewell, W. *Philosophy of the Inductive Sciences, founded upon their history.* 2 vols. London, 1847.

White, Andrew D. *Seven great statesmen in the warfare of humanity with unreason.* New York, 1910.

White, Andrew D. *A history of the warfare of science with theology in Christendom.* 2 vols. New York, 1896.

Wilde, Emil. *Geschichte der Optik, vom Ursprunge dieser Wissenschaft bis auf die gegenwärtige Zeit.* 2 vols. Berlin, 1838–43.

Wolf, G. *Einführung in das Studium der neureren Geschichte.* Berlin, 1910.

Wolf, R. *Geschichte der Astronomie.* Munich, 1884.

Wood, Anthony. See Clark.

Wordsworth, Christopher. *Scholae academicae: Some account of the studies at the English universities in the eighteenth century.* Cambridge, 1877.

Zedler, J. H. *Grosses vollständiges Universallexicon aller Wissenschaften und Künste.* . . . . Halle und Leipzig, 1732–50.

Zittel, K. A. *Grundzüge der Palaeontologie (palaeo-zoologie).* München, 1885.

Zöller, Egon. *Die Universitäten und technischen Hochschulen. Ihre Geschichte Entwicklung und ihre Bedeutung in der Kultur, ihre gegenseitige Stellung und weitere Ausbildung.* Berlin, 1891.

INDEX

INDEX

Aberdeen, University of, 235
Academia Caesarea Leopoldini; *see* Collegium Naturae Curiosorum
Academia Constantium in Padua, 69
Academia Investigantium in Naples, 69
Academia Naturae Curiosorum; *see* Collegium Naturae Curiosorum
Academia Philoexoticorum in Brescia, 69
Academia Physico-critica in Sienna, 69
Academia Physico-Mathematica in Rome, 69
Academia Secretorum Naturae, 74; *see* Della Porta
Academicus, title used by Galileo, 75
Académie des Sciences, 139–64; transfusion of blood at, 118; had support of Louis XIV, 139; beginnings of the, 139; Mersenne and his associates, 140–43; meetings at Montmort's, 143, 144; at Thevenot's, 144; Colbert's organization of, and list of appointees, 144–47; joint meetings of physicists and mathematicians, 147; *pensionnaires*, 148; method and line of work, 148–49; *L'Histoire naturel des animaux* published, 149–50; *L'Histoire des plantes*, 150; Paris Observatory established, 151; scientific expeditions of the, 152; early experiments by members, 153–54; Huygens' *Horologium* published by, 154; artisans' methods and instruments investigated, 155; map of France, 156; personnel of, changed, 156; decline of, on death of Colbert, 145, 156; more practical work demanded by Louvois, 157; Cartesians and anti-Cartesians, 157; visited by James II, 158; Lémery and Homberg members, 158; new members, 158 (n. 71); 159 (n. 74); reorganized under Bignon, 159; new statutes of, 160–61; Fontenelle on the work of, 162; influence of, on progress of science, 162–64; the model for other academies, 164

Académie des Sciences: and the Royal Society, 91–92, 139, 144, 163; correspondence of Royal Society with, 104; *History of Animals* issued by, translated at instance of Royal Society, 129, 150
Academy of St. Petersburg, due to Leibniz's suggestions, 196
Accademia Curiosorum Hominum, 73–74; *see* Della Porta
Accademia dei Lincei, 68; forerunner of the Cimento, 74; formed by Cesi, 74; members, 74; rule of studious life of, 75; *Gesta Lynceorum*, 75; Galileo's affiliations with, 75; publications of the, 75–76; ceased to exist, 76; connection of Cimento with, through Galileo, 76
Accademia del Cimento, 68; first organized scientific academy, 73; its antecedents, 73–74; its founders, 76–77; members of the, 78–81; corresponding members, 81–82; work and method of, 82–83; researches, described in the *Saggi*, 83–88; translations and spread of *Saggi*, 88; the *Atti e memorie*, 89; Rosenberger on the, 89–90; English and French societies had much in common with the, 92
Accademia del Principe, 78
Acoustics, knowledge of, in 1600 and in 1700, 7
Acquaviva, *Ratio studiorum*, 215
Acta eruditorum: edited by Otto Mencke, 203–4; folios in Latin, 204; purpose and contents of, 204–6; made known German scientific thought, 206
Acta medica et philosophica hafniensia, edited by Bartholinus, 203
Advance, general scientific, in seventeenth century, 3–20
Aerskin, William, charter member of Royal Society, 105
Air pump: Guericke and Boyle on, 7; invented by Guericke, 50–52;

288 THE RÔLE OF SCIENTIFIC SOCIETIES

Boyle's experiments with the, 59; Fontenelle marvels at the, 162

Aix, a society at, 69

Alchemy, laboratories of, 4

Allen, Thomas, at Worcester College, 97 (n. 22)

Altdorf, the University of: most progressive, 175, 180; Leibniz took degree at, 180; conspicuous for cultivation of science, 230, 235; established well-equipped chemical laboratory, 231 (n. 63); anatomical theater at, 232

Amateurs in science, 55–66; Boyle, 57–61; Lord Bruce, 61; F. Cesi, 55; Charles II, 57; Crabtree, 63; K. Digby, 62; Evelyn, 61; the Fuggers, 64; Guericke, 65; Sir Mathew Hale, 61; Hartlib, 62; Hevelius, 65; Abraham Hill, 63; J. Horrox, 63; Huygens, 64; Leibniz, 65; Leuwenhoeck, 64; Marsiglio, 55; F. and L. Medici, 56; W. Molineux, 63; Sir R. Moray, 61; J. Moritz, 63; Duke of Orleans, 56; Peiresc, 56; W. Petty, 62; F. Potter, 62; Prince Rupert, 57; Tschirnhausen, 65; Van Helmont, 64; Lord Willoughby, 61; Marquis of Worcester, 61

Amsterdam, toleration in, 251

Anatomical instruction: the verbal analysis of Galen, 215; reform necessary, 215–16; in Padua, 217; in Oxford, 236, 242; reform adopted, 255

Anatomical theaters at several universities, 232, 255

Anatomy, comparative: born, 13; Grew, founder of, 133; Fontenelle on, 162; *see* Dissection

Andreae, Valentin, suggested founding an academy of natural sciences, 168

Antiperistasis, Aristotelian idea of, disproved, 87

Apollonius, works of, reconstructed by Viviani, 79

Aristocracy, English, interested in science, 57

Aristotelian ideas: of dynamics, 6; of astronomy, 9; of chemistry, 9; opposed by Gilbert, 23; disproved by Galileo, 23–24; refuted in Galileo's *Dialogo*, 28–29; errors in, refuted by Renaldini, 80; thesis opposing the, 221–22

Aristotle: impossible to be excelled, 131; Glanvill's *Progress . . . . since*, 131; domination of, in universities, 213–15, 220–21, 237, 238; opposition to domination of, of Galileo, 29–30; of Leibniz, 179–83; of Thomasius, 234

Arithmetic, knowledge of, in 1600 and in 1700, 17

Artisans, methods and instruments of, investigated by Académie des Sciences, 155

Aselli: discoverer of the lymphatics, 38; taught at Pavia, 219

Ashmolean Museum, the, at Oxford, 247

Astronomy: knowledge of, in 1600 and in 1700, 8–9; study of, based on Ptolemy's treatise, 213; condemnation of Galileo a check to study of, 218

Atmospheric pressure, 7; Torricelli on, 32–33; Pascal and Périer on, 33; experiments of the Cimento on, 84–86; *see* Boyle, Guericke

Aubrey, John: on Harvey, 37; on Wilkins, 96; on Rooke, 97

Auzout: asked Louis XIV for an observatory, 151; perfecting the telescope, 153; practical program of, 155

Bacon, Sir Francis: iconoclast and builder, 39; message of, 40; the "instances" of, 41–42; on compiling "histories," 41–42; through learned societies, 42–43; the "House of Salomon," 43, 264–70; hated erudition, 45, 258 (n. 5); agreement of Descartes with, 44–46; works of, widely read, 53; Royal Society followed program sketched by, 92; criticism of the English universities, 239–41, 258 (n. 5); dedicated books to Cambridge, 240; ideas of, cherished at Cambridge, 247; not affiliated with a university, 257

Bacon, Roger: knowledge of optics of, 7; held university system valueless, 257–58

Ball, William: charter member of Royal Society, 100, 105; Society met at chambers of, 101; treasurer, 105

Barnacles, Moray on, 104

Barometer, the, discovered by Torricelli, 32–33

INDEX

Barrow, Isaac: Fellow of Royal Society, 107; first holder of Lucasian professorship, 248

Bartholin, Erasmus, on Icelandic spar, 15, 154

Bartholinus, Thomas, editor of *Acta medica*, 203

Basnage de Beauval, Henri, editor of *Histoire des ouvrages des savants*, 208

Bathurst, Dr. Ralph: Oxford member of Royal Society, 96; at Worcester College, 97, 99

Bauhin, on classification in botany, 12

Bausch, Dr. Lorentz: founder of the Collegium Naturae Curiosorum, 169; death of, 170; papers by, published, 171, 201

Bayle, Pierre: praised the *Miscellanea curiosa*, 174; published a periodical, 203; *Nouvelles de la république des lettres*, 207–8

Beal, Dr., letter of, to Boyle, on interest of Oxford students in Royal Society, 246

Berlin Academy, the: 68; due to Leibniz, 177; aided by plan of Electress to build observatory, 190; outcome of a plan of calendar reform, 190–91; program for, 191; charter and statutes, 192; dragged along for ten years, 192; activity of Leibniz for, 193; financial difficulties, 193; first volume of *Miscellanea* published, 194; ill treatment of Leibniz and decline, 194

Bernard, Dr. E., on the Royal Society, 260 (n. 11)

Bibliography, 271–83

Bibliothèque universelle et historique, by Leclerc, 208

Bignon, Jean Paul: reorganized the Académie, 159; on the task of the Académie, 161

Binot and Couplet, catalogue of devices, 155

Blegny, Nicholas de, publishes *Nouvelles descouvertes sur toutes les parties de la médecine*, 203

Blondel: described machines to the Académie des Sciences, 155; applying mechanics to artillery, 156

Blood, circulation of the: Galen's notions, 13–14; Cesalpinus and Fabricius on the, 34; Harvey on the, 34–37; Malpighi's, Aselli's, Pecquet's proofs of the, 38

Blood, transfusion of, 62; experiments of Royal Society, 119; of Académie des Sciences, 153–54; first tried by a German, 185; symposium on, in *Journal des sçavans*, 201

Blood corpuscles, red, Leuwenhoeck's paper on, 130

Bologna, experimental academy of, founded by Duke Marsiglio, 69

Bologna, University of: rich and independent, 217; German students at, 217; Marsiglio bequeathed laboratories to, 219; no change in laws of theological faculty of, 259

Borelli: an iatrophysicist, 15; most famous member of the Cimento, 79; quarrels of, with Viviani, 79; experiments of, showing air necessary to life, 86; defense of Copernican system by, 89

Botanical gardens: multiplication of, 5; necessity for, 216; at University of Paris, 225; at most universities, 231, 254

Botany, knowledge of, in 1600 and in 1700, 11–12

Bourdelin: joint author of *Histoire des plantes*, 150; on Vichy water, 154

Boyle, Robert: laws of elasticity of gases tested by, 5; on air pump, 7; on chemistry, 11; on crystallization of bismuth, 15; "touched" by Valentine Greatrix, 20; and the "Christian epic," 20; not a generalizer, 52; foremost of English scientists, 57; on science and religion, 58; air-pump experiments of, 59, 103; the *Sceptical Chymist* of, 59; on right methods of experimentation, 60; co-operated with societies, 68; correspondent of the Cimento, 82; "Invisible College" of, 95 (n. 12); Oxford meetings at lodgings of, 96; Rooke, assistant of, 97; brought Peter Stahl to Oxford to teach chemistry, 99; Evelyn's letter to, 99 (n. 30); important rôle at organization of Royal Society, 103; questions sent to Teneriff, 103 (n. 46); charter member of Royal Society, 105; friend of Oldenburg, 106 (n. 52);

released Hooke to Royal Society, 108, 111; on spontaneous generation, 117; vacuum experiments of, shown and tested, 119, 123 (n. 132), 201; objects to Royal Society founding a college, 122; Stubbe's correspondence with, 132; Tuke related experiment of, to Académie des Sciences, 143; need of communication with other scientists, 199; denounced by the Oxford pulpit, 246; not affiliated with a university, 257; on the Royal Society, 260 (n. 11)

Boyle's law, discovered by one of his helpers, 52

Breathing, the process of: Galen's view, 13; parallel with burning, 15, 117

Breréton, William, charter member of Royal Society, 100, 105

Briggs, on logarithms, 18

Brooks, Lord, endowed lectureship in history at Cambridge, 237 (n. 93)

Brouncker, William, first president of the Royal Society, 100, 105

Bruce, Lord, an amateur of science, 61

Buono, Candido and Paolo del, members of the Cimento, 80

Calculating machine, made by Leibniz, 186

Calculus, infinitesimal, victory of the, in Académie des Sciences, 157

Calendar reform, the Berlin Academy begun in a plan for 190–91

Cambridge University, 235; Brooks's lectureship in history at, 237 (n. 93); Bacon's comments on, 239; Bacon dedicated books to, 240; lowest level of science at, 240; no astronomy or mathematics at, 241; invective of Dell against, 242; the "new philosophy" at, 247; Cartesianism at, 248; Barrow and Newton in Lucasian professorship of science, 248; how the Newtonian philosophy took root at, 249

Camera obscura, described by Della Porta, 7

Camerarius, on sex in plants, 13

Campanella, held university system valueless, 258

Camus, prohibited from lecturing in French, 221

Capillaries, Malpighi on the, 38

Carcavi, charter member of Académie des Sciences, 172

Cardan, held university system valueless, 258

Carrichter, classification of flowers by, 12

Cartesianism: a religion, 50; first Académie des Sciences ignores, 147 (n. 28); later battleground between, and anti-Cartesians, 157; attitude of University of Paris to, 221–23; welcomed by Cambridge Platonists, 247–48; Newton's deathblow to, 289; in Dutch universities, 253; attitude of universities, 255; *see* Descartes

Cassini, Giovanni Domenico: a guest of the Cimento, 82; on the meetings in Mersenne's cell, 142; made head of Paris Observatory, 152; optical instruments and models of, 155; deciphered a Siamese astronomy, 158

Cavalieri: on the use of infinitesimals, 18; corresponded with Descartes, 140

Cayenne, scientific expedition to, 152

Censorship of books: Leibniz and, 183; function of University of Paris, most strict, 221, 255; non-existent in Holland, 251; in England, 255

Cesalpino: botanical ideas of, 12; on circulation of the blood, 34

Cesi, Frederico: a lover of science, 55; founder of the Accademia dei Lincei, 74; death of, 76

Charles II: persons touched by, 20; an amateur in science, 57; discussed scientific questions with Brouncker and Evelyn, 100; approved Royal Society, 102; gave the charter, 104–5; ordered patents to be approved by Royal Society, 121

Charter of the Royal Society, 104–5

Chauvin, Etienne, editor of *Nouveau journal des savants dressé à Rotterdam*, 208

Chemistry: experimentation in, 4; knowledge of, in 1600 and in 1700, 8–12; Van Helmont, pioneer in, 38; amateurs of, 55, 56; Boyle's experiments in, 59 60; taught privately at Oxford, 98, 245; in Académie des Sciences, 151, 158, 159; interdicted at

INDEX

University of Paris, 224; first professor of, in Marburg, 230; fostered at University of Altdorf, 231; neglected at Erfurt, 231–32; interest in, at Oxford, 242, 244, 247; Hall complains of lack of, 242

Christian epic, the, adherence of scientists to, 20

Clark, Dr. Timothy: charter member of Royal Society, 100, 105; injection of liquid into veins of animals, 103

Clarke, Dr. Samuel, translated Rohault's *Physics* into Latin, 249

Clinical teaching: at Padua, 218; at polyclinic in Paris, 225 (n. 36); at Leyden University, 252–53, 255

Coga, Arthur, sheep's blood transfused into, 119

Colbert, Jean Baptiste: an observer of experimenters, 56; organized the Académie des Sciences, 144-45; appointees of, 145–46; death of, 156; permits publication of *Journal des sçavans*, 199–200

Collectanea medico-physica, published in Dutch, 203

College, the Invisible, of Boyle, 95 (n. 12), 106, 241

College: question of forming a, by the Royal Society, 122; plans of Wren for a, 122 (n. 127); opposed by Boyle, 122

Collège de France: founded by Francis I, as a home for humanism, 225–26; Ramus, Gassendi, and Roberval at, 225–26

Collegium Curiosum sive Experimentale, 175–77; established by Christopher Sturm, 175–76; express purpose of, 175; two volumes of experiments published by the, 176–77; trained clever experimenters, 177

Collegium Naturae Curiosorum, 169–75; not a scientific society, 169; union of physicians publishing a scientific periodical, 169; founded by Dr. Bausch, 169–70; extended by Dr. Sachs, 170; statutes, 171; *Miscellanea curiosa* of, 171–73; patronage of emperor, 173; changes of name, 173; obstacles to, 173–74; comments on the *Miscellanea*, 174; features of work of, 174–75

Collins, J., voluminous correspondence of, 198

Colonna, Fabius: member of the Lincei, 74; botanical observations of, published, 76

Colors: Newton's theory of, presented to Royal Society, 135–36; criticized by Hooke and Huygens, 136

Columbus, Rualdus, pupil of Vesalius, 34

Colwall, David, started rarity cabinet of the Royal Society, 114

Comets: Tycho Brahe on, 8; number of *Journal* devoted to, 201–2

Congregation of the Index, the, condemned the Copernican system and Galileo's teachings, 28, 256

Conservatism of the university system: Leibniz's objection to, 178, 180, 181, 182; in Padua, 218; in Paris, 220–22; in Germany, 229–30; opposition to, of Thomasius, 232–34; in Oxford, 237; in Dublin, 238; in Glasgow, 238; Bacon on, 239–40; criticism of, by Hall, 242; by Webster, 243; sarcasm about opponents to, 246–47; less at Cambridge, 247–48; introduction of Newtonian philosophy illustrates, 249, 250–51; least in Holland, 253; instances of, 258

Copernican system: the work of the sixteenth century, 8; proofs of the, 9; Galileo's proof of the, 26–27; attacked by the Dominicans, 27; condemned by the Congregation of the Index, 28; discussed in Galileo's *Dialogo*, 29–31; defense of, by Borelli, 89; study of, in program of Royal Society, 95; Galileo did not lecture on, 218; not condemned by Paris faculty, 223; mention of, enjoined upon Savilian professor, 237; taught in Oxford, 244; attitude of the universities toward the, 255

Correspondence: with foreign learned bodies opened by the Royal Society, 104, 123–24; unreliability of personal, 199

Cosimo II made Galileo court mathematician and professor at Pisa, 27

Couplet, Binot and, catalogue of mechanisms, 155

Crabtree, an amateur astronomer, 63

Crew, Nathaniel, pupil of Peter Stahl, 99
Croone, Dr.: member of Royal Society, 100; widow of, established a lectureship, 122
Crosse, Rev. Robert, attack of, on Royal Society, 131
Crystallography, knowledge of, 15
Curator, of the Royal Society, 109; see Hooke
Cutler, Sir John, initiated lectures by the Royal Society, 122

De la Boe, an iatrochemist, 15
De la Hire; see La Hire, Philippe de
Degree, Bachelor's, courses leading to the, 213
Degree, Master's, advanced studies for the, 213
Dell, William, invective of, against Cambridge, 241
Della Porta, Johann Baptista (see also Giambattista): *Magiae naturalis* of, of, 7; academy of, 73–74; member of Accademia dei Lincei, 74
Dépêche du Parnasse ou la gazette des savants, 208
Desargues, Gaspard, founder of descriptive geometry, 141
Descartes: on lenses 7; on vortex motions, 9, 48, 49; on the eye, 14–15; the iatrophysicists followed, 15, 48; on algebra and geometry, 18, 19; significance of, for the cause of experimental science, 44; on experiments, 45; hated mere erudition, 45–46, 258 (n. 5); wrote in the vernacular, 45; applies reason to question of divinity, 46; laws of, on the mode of reasoning, 47; the scientist, 47–48; timidity of, in expressing his scientific conviction, 48–49; "*einheitliche Weltanschauung*" of, 49; Boyle opposed to, 57–58; correspondents of, through Mersenne, 140; Cartesian views of Montmort's gatherings, 143; *Méthode* of, dedicated to the Sorbonne, 222; works of, put on the Index, 222; opposition to, formulated, 223; *Méthode* of, placed beside Aristotle's *Organon* at Paris University, 256; not affiliated with a university, 257; see Cartesianism

Digby, Kenelm: an amateur of science, 62; charter member of Royal Society, 105
Disputations, a main feature in university instruction: character and subjects of, 214; in medical course at Oxford, 236; at Erfurt, 229; perpetuated at Halle, 254
Dissection, methods of: adopted, 5; of animals, 12; Vesalius insisted on, of man, 34; Harvey tireless at, 35–37; Descartes' advice on, 45; Boyle on, 60; a "fad," 65; performed by Stenon, 143, 144; in Académie des Sciences, of animals, 149; human, 153; Leibniz desired, as part of school curriculum, 181; in German scientific society, 184; Vesalius adopts, in University of Padua, 218; Massari in Bologna, 219; in Paris, 221; at German universities, 231; at Altdorf, 232 (n. 70); in Würzburg, 232
Divini: controversy of, with Huygens, 89; articles by, in *Journal*, 201
Dodart, Denis: joint author of *Histoire des plantes*, 150; chemical studies of plants of, 151
Dominicans, the, attack the Copernican system, 27
Dresden Academy, plan of, by Leibniz, accepted by ruler of Saxony, 195
Dublin University, 235; adopted Laudian Statutes, 238; studies at, 238
Duclos: studied the flow of sap, 151; collaborated on *Histoire des plantes*, 150; on Vichy waters, 154
Duhamel, Jean Baptiste: secretary of Académie des Sciences, 145 (n. 25); wrote history of Académie, 142 (n. 8), 144, 148
Dynamics: knowledge of, in 1600 and in 1700, 6–7; Galileo founder of modern science of, 23–26; problems in, investigated by Royal Society, 117

Eckius, Johannes, member of the Lincei, 74
Edict of Nantes, revocation of, affects Académie des Sciences, 156
Edinburgh, University of, 235; Newton's ideas introduced in, by David Gregory, 250; professorship of botany at, 250

INDEX   293

Education, reform of: interests Ratichius, Jungius, 167; Weigel, 180; Leibniz, 181, 184; Leibniz's suggestions for Russia, 195; opposed by University of Paris, 223-24

Electress, the, planned an observatory at Berlin, 190

Electricity: knowledge of, in 1600 and in 1700, 8; Gilbert the father of, 23

Elements: the four, of Aristotle, 9; the three, of Paracelsus, 10; the two, of Van Helmont, 38

Elizabethan statutes of 1570, 235

England: amateurs in, 57-63; Royal Scientific Society of, 91-138; universities of; see English universities

English universities: affected by political disturbances and religious wranglings, 235; general features of, same as in Europe, 235; interest in science at lowest ebb, 236; medical instruction at Oxford, 236; Laudian statutes, 237; Bacon on the, 239; lowest level at Cambridge, 240; remarkable men at Oxford, 241; criticisms of the, 242-44; reply of Wilkins and Ward, 244-45; science at Oxford, 245-46; the "new philosophy" at Cambridge, 247; the Newtonian philosophy introduced, 254

Ent, Dr. George: charter member of Royal Society, 93, 100, 105; friend of Harvey, 95; on air in burning and breathing, 118; on transfusion of blood, 119

Erfurt, University of: disputations required at, 229; examination for medical degree at, 232

Erlangen, professor of chemistry at, had to have laboratory in his own house, 259

Eulenburg, Franz: on the German universities, 226-29; and the academies, 262

Evelyn, John: a amateur of science, 61; on Wilkins, 96; on a philosophic-mathematical college, 99 (n. 30); charter member of the Royal Society, 100, 105; *Sylvia* of, published by the Royal Society, 129

Expeditions, scientific, of the Académie des Sciences, 152

Experimentation: introduced into science, 4-6, 21, 22; produced scientists, 5, 18-19; pioneers of, Gilbert, 22-23; Galileo, 24-28; Torricelli, Pascal, 33; Harvey, 35; only method, 38; of Van Helmont, 38; Bacon, promoter of, 39-40; purpose of Bacon's "House of Salomon," 43, 264-70; Descartes' attitude toward, 45, 46, 47, 49; of Guericke, 51-52; worship of, 52; how knowledge of, spread, 52-53; the class interested in, 53-54; in competition with scholastic learning, 54; amateurs interested in, 54-64; Boyle on usefulness of, 58; on right methods of, 60; popular work on, 65; enthusiasm for, led to formal affiliation, 67; called forth societies, 67; in Della Porta's Academy, 74; efforts of Cimento concentrated on, 82-88, 90; by members of Royal Society, 93, 94, 95, 98, 112, 116, 131; how carried on in Royal Society, 103; statutes of Royal Society on, 108-10; instruments of Royal Society for, 112-13; sketch of, done by Royal Society, 116-19; methods of, in Académie des Sciences, 148; sketch of, done by Académie des Sciences, 149-50, 153-54; Louvois objects to, 157; dearth of, in Académie des Sciences, 157-58; of Lémery, 158; Fontenelle on, 162; Jungius on, 167; medical science rests on, 174; express purpose of Collegium Curiosum sive Experimentale, 175-77; of Leibniz, 179-80, 181, 186; importance of, as part of education, 185; Leibniz plans societies for, 184, 187, 188, 191; of Berlin Academy, 194; century of, gives rise to scientific journals, 198; must be introduced in universities, 216; at Italian universities, 217-18; opposition to, at University of Paris, 223-24; interest in, at German universities, 229-32; cherished at Oxford "extra Collegium," 241, 243; criticism of lack of, at Oxford and reply, 243, 244; beginning of, at English universities, 248, 249-50; Newton tireless at, 248; at University of Leyden, 251-52; progress of, in universities, 252; fostered in scientific societies, not in universities, 259-61; *see* Instruments, Laboratories

Fabri, Honoré, Jesuit, correspondent of the Cimento, 81–82
Fabricius: on the valves of the heart, 34; Harvey a pupil of, 34
Faculties: must be changed, 216; relative attendance of students in, of German universities, 228 (n. 50); no change in, 255
Falling bodies; see Dynamics
Falloppio, on circulation of the blood, 34
Fellows of the Royal Society: named, 107; personnel of the, 110–11
Fermat, Paul, on the theory of probability, 18, 141
Filiis, Anastasio de, member of the Lincei, 74
Fischer, Kuno, on Leibniz, 178, 180
Flamsteed, John: member of Royal Society, 132; astronomer at Greenwich, 133; not affiliated with a university, 257
Flying, artifices of, shown by Hooke to Wilkins, 108
Fontenelle, Bertrand le Bovier, on the work of the academicians, 162
Fossils, explanations of, 16
Foster, Samuel: at earliest meetings of Royal Society, 93; professor of astronomy at Gresham College, 94
France, the *Kulturträger* of the Continent, 164; amateurs of, 56; scientific societies of, 139–64; scientific journals of, 198–202, 203, universities of, 220–26
Francis I, founded Collège de France, 225
Frederick I, of Prussia, Elector: marriage of, 189; added study of the German language to program of Berlin Academy, 191; granted Thomasius *venia legendi* in Halle, 234
Frederick II, the Great, on Leibniz, 178
Freedom of conscience: must obtain in the university, 216; only in the North Italian institutions, 217, 255
Freedom of speech in Amsterdam, 251
Freedom of the press: there must be, 216; in Amsterdam, 251; in Holland and England, 255
Freedom of thought: there must be, 216; suppression of, 221–24; limits placed on, 255
Freiburg, University of: relative attendance in faculties of, 228 (n. 50); book of Peter Ramus forbidden at, 228; instruction at, 266; anatomical theater at, 231
Fruchtbringende Gesellschaft, the, 166, 182 (n. 72), 185; appeal of Leibniz to the, 188
Fuggers, the, took L'Ecluse on their travels, 64–65

Galen: teachings of physiology of, 13–14; omniscience of, refuted, 34; medical instruction a commentary on, 214
Galileo: on dynamics, 6–7; on cycloid curves, 18; basis of popular fame of, 23–24; the experimenter, disproving Aristotle's notions on falling bodies, 24–26; discovery of the telescope, 26; *Sidereus nuntius* of, 27; on the Copernican doctrine, 27, 28; made court mathematician and professor at Pisa, 27; contest of, about phases of Venus, with Jesuit Scheiner, 27; analysis of his *Dialogo*, 28–31; results of condemnation of, 31–32, 76, 219; strangely conservative, 32; member of the Lincei, 74; close affiliation of, with the Lincei, 75; books of, published by the Lincei, 75; spiritual father of the Cimento, 77; the Medici brothers pupils of, 77; air thermometer of, 83; experiments of, on falling bodies proved by the Cimento, 87–88; work of, studied by Royal Society, 91, 117; corresponded with Descartes through Mersenne, 140; works of, translated by Mersenne, 140; expounded the Ptolemaic system at University of Padua, 218
Gallois, Abbé: joint author of *Histoire des plantes*, 150; opposed the Newtonian calculus, 157; editor of *Journal des sçavans*, 153, 200
Gassendi: on Peiresc, friend of scientists, 56; on velocity of sound, 88; corresponded with Descartes, 140; associate of Mersenne, 142; teachings of, forbidden, 223; defended the Copernican hypothesis, 223; professor at Collège de France, 226, 227

INDEX 295

Generation, spontaneous: denied by Redi, 80; studies in, 117
Geology, knowledge of, in 1600 and in 1700, 15–16
Geometry, knowledge of, in 1600 and in 1700, 7–18
German language; *see* Vernacular
Germany: amateurs in, 64–65; science backward in, 165; slow development of the vernacular in, 165; the Sprachgesellschaften, 165; scientific societies of, 166–97; the Societas Ereunetica of, 167–68; the Collegium Naturae Curiosorum of, 169–75; the Collegium Curiosum sive Experimentale of, 175–77; Leibniz plans society for, 184–85, 185–86, 187–88, 189, 189–90; Berlin Academy of, 190–95; no place for experimental science in, 197; scientific journals of, 203–6; the *Acta eruditorum* of, 203–6; Franz Eulenburg on the universities of, 226–29; Protestant and Catholic, 227; reasons for founding, 226–27; religious element dominant in, 227–28; type of work in, 228–29; science at Altdorf, 230–31; Thomasius and establishment of the vernacular, 232–33
Gesner, Conrad, on plants, 12
Gesta Lynceorum, the earliest scientific-society publication, 75
Gilbert, William: father of electricity, 23; first to use experiment only, 22; Galileo on the discovery of, 31; errors in, refuted by Renaldini, 80–81
Giornale de' litterati di Roma: edited by Ricci, 81, 202; extracts from, in *Journal des sçavans*, 202
Glanvill, Joseph: on Bacon's "House of Salomon," 43; on the telescope, 53; defense of the Royal Society, 131
Glanvill, William, poem of, on the Royal Society, 102
Glasgow University, 235; studies at, 238; rejected Cartesianism, 248; obtained telescope from George Sinclair, 250
Glisson, Dr.: at earliest meetings of Royal Society, 93; friend of Harvey, 95
Goddard, Dr. Jonathan: at earliest meetings of Royal Society, 93; Cromwell's physician, 94; drudge and chemist of the Society, 94; charter member, 105; on worms in vinegar, 117
Goethe, on Jungius, 169
Göttingen, University of, an ideal home for science, 259
Graaf, Reinhart de: experiments of, tested, 119–20; on the microscopes of Leuwenhoeck, 129
Gregory, David: introduced discussion of Newton's ideas at St. Andrews and Edinburgh, 250; on the Royal Society, 260 (n. 11)
Gresham, Sir Thomas, founder of Gresham College, 100
Gresham College: founded, 100; scientific meetings at, 101; Royal Society organized at, 101–2, 235 (n. 83); place of meeting, 102, 110 (n. 69); a slash at, 247; Hooke professor at, 257
Grew, Nehemiah: microscopic researches of, 13; on sex in plants, 13; list of Royal Society's instruments, 114–15; member of Royal Society, 132; works of, 133; articles on, in *Journal*, 201; studied at Leyden, 253; not affiliated with a university, 257
Grimaldi, a Jesuit: on diffraction, 7; skilful experimenter, 218
Groningen, University of: German and English students at, 251; Cartesian teaching overthrew Aristotelian physics at, 253
Guericke: on air pump, 7, 50–52; on electricity, 8; continuator of Galileo's and Torricelli's work, 50; experiments of, 51–52; took up engineering, 65; spent $20,000 on instruments, 67; experiments of, tested, 119–20; shown at Würzburg, 230; studied at Leyden, 253; not affiliated with a university, 257
Gulielimini, on crystals, 15
Gunpowder in cylinder, Huygens' and Papin's experiments on, 153

Hafnia, a learned society at, 69
Hale, Sir Mathew, an amateur of science, 61
Hall, John, criticism of, on the universities, 242

Halle, University of: question of witch trials discussed at, 20; foundation of the, 234; the vernacular at, 234, 255; *libertas philosophandi* adopted at, 256

Halley, Edmund: on motion of celestial bodies, 9; on comets, 9; on mortality tables, 18; member of Royal Society, 132; Fellow, work of, 134; Savilian professor of geometry at Oxford, 134, 257 (n. 2); consulted Newton on path of the earth, 137; published Newton's *Principia* at own expense, 138

Hank, Theodore: first suggested meetings which led to foundation of Royal Society, 93, 95; F.R.S., 107

Hannover, court of, Leibniz librarian and confident at, 188

Harnack, Adolph: on Leibniz, 178, 179; on the program for the Berlin Academy, 191; on universities and academies, 261

Hartlib, Samuel: an amateur of science, 62; corresponded with Descartes, 140

Hartmann, Johann, first professor of chemistry in Europe, at Marburg, 230

Hartsoecker, associate member of the Académie des Sciences, 156

Harvey: on circulation of the blood, 14, 35–39; a pioneer in medicine, 34; Aubrey on, 37; work of, studied by Royal Society, 92

Heidelberg, University of: greatest liberality at, 228; number of professors at, 229

Helmstadt, University of, founded by Julius of Brunswick, 227

Henry IV, gave new statutes to University of Paris, 220

Henry VIII, endowed Regius professorship at Oxford, 236

Henshaw, Thomas, charter member of Royal Society, 100, 105

Heretics, excluded from candidature at University of Paris, 224; *see* Theology

Hevelius: on comets, 9; built a private observatory, 65; correspondence with, 124; articles by, in *Journal*, 201; studied at Leyden, 253; not affiliated with a university, 257

Hill, Abraham, an amateur of science, 63, 100

Hippocrates, medical instruction a commentary on, 214

Histoire des ouvrages des savants, by H. Basnage de Bouval, 208

Histoire des plantes, by Duclos and others, 150

Histology, foundation of, laid, 13

Histories, Bacon on compiling, 41–42

Hobbes, Thomas: corresponded with Descartes, 140; criticism of universities in his *Leviathan*, 243, 258 (n. 5)

Hoffman, Moritz, built the first chemical university-laboratory at Altdorf, 231

Holland: amateurs of, 63; periodicals published in, 207–8; university development in, 250; universities founded by wealthy cities, and scholarship encouraged, 250–51; Leyden, 251–53; no liberty of religious belief, 251; medicine and clinical teaching at Leyden, 252–53; Cartesian teaching in, 253

Homberg, Wilhelm: chemist of Duke of Orleans, 158–59; director of chemical laboratories of Académie des Sciences, 159

Hooke, Robert: on light, 7; on motion of celestial bodies, 9; on fossils, 16; on experimentation, 52; on wonders of the microscope, 53; F.R.S., 107; appointed curator of Royal Society, 107; exhibited experiments at every meeting, 116; on micro-organisms, 116–17; on air essential to life, 117; *Micrographia* of, published by Royal Society, 129; rejected Newton's theory of colors, 136; on falling bodies and motion of the earth, 136–37; claims priority of Newton, 137–80; quarrel of, with Huygens, 199; *Micrographia* of, reviewed in *Journal*, 201; other publications of, 208; professor at Gresham, 257

Horrox, Jeremia: an amateur astronomer, 63; on lack of astronomy at Cambridge, 241

"House of Salomon," Bacon's, 43–44; a forerunner of, 75; text of, 264–68

Humanistic studies not cultivated by the Lincei, 75; *see* Vernacular

Huxley, on the *Philosophical Transactions*, 125

INDEX   297

Huygens, Christian: on light, 7; on comets, 9; on cycloid curves, 18; on the use of infinitesimals, 18; gave his whole life to science, 64; co-operated with societies, 68; correspondent of the Cimento, 82; controversy of, with Divini, 89; gave lens to Royal Society, 114; experiments of, tested, 143; correspondence with, 119; opposed Newton's theory of light and colors, 136; corresponded with Descartes, 140; letter of, on rings of Saturn, read to Montmort's Academy, 143; member of Académie des Sciences, 145 (n. 25); plan of work for the Académie, 149 (n. 37); the experiments of, 153; undulatory theory of light of, 154; *Horologium* of, published, 154; mechanisms of, in catalogue, 155; lost to the Académie, 156; opportunity given to, 163; Leibniz a correspondent of, 179; need of communication with other scientists, 199; quarrel of, with Hooke, 199; contributed to the *Journal des sçavans*, 201; pendulum clock of, in *Journal des sçavans*, 201; criticism of Newton's telescope, 201; willed his MSS to Leyden, 253; not affiliated with a university, 257

Hydrostatics: knowledge of, in 1600 and in 1700, 7; laws of, discovered, 33

Hygrometer, of Ferdinand de Medici, 84

Iatrochemists, the, 15; Van Helmont, 39; De la Boe, 252

Iatrophysicists, the, 15; Descartes, 48

Innsbruck, University of, refused to establish professorship in botany and chemistry, 259

"Instances," the, of Francis Bacon, 41–42

Instruction, mode of: reform needed in, 215; no change in, 254

Instruments, scientific: creation of seventeenth century, 4, 21; developed by Galileo, 23; expense of, 67; of dukes of Medici at disposal of Cimento, 78; measuring, studied by Cimento, 83–86, 89; best made by Italians, 90; list of, of Royal Society, by Sprat, 112–14; by Grew, 114–15; discussed in *Philosophical Transactions*, 127; collection of, in Académie des Sciences, 155; of Sturm, 176; of Berlin Academy, 191; at University of Altdorf, 230–31; Bacon's commentary on, 240 (n. 106); lack of, at Oxford, 246; requested for University of Glasgow, 250; made possible by scientific societies, 260; in Bacon's "House of Salomon," 264; see Experimentation, Laboratories, Air pump, Barometer, Microscope, Thermometer, Telescope

Invisible College, an, at Oxford, 95 (n. 12), 241, 244

Italy: amateurs of, 55–56; scientific societies of, 68, 69, 73–90, 219; journal published in, 202; universities of, 217–19; Padua, Pisa, and Bologna, 217; pioneers in experimental study of medicine, 217; Vesalius at Padua, 218; experimental physics, 218–19; laboratory of Marsiglio given to Bologna, 219; progressive features of, 253–55

Jablonski: suggested an observatory to the Electress, 190, 191–92; ill treatment of Leibniz by, 194

Jablonski brothers, friction of Leibniz with the, 193

James II, visited the Paris Observatory, 158

Java, queries and replies from, 121

Jena, anatomical theater at, 231

Jesuits, Scheiner: opponent of Galileo, belonged to, 27; correspondent of Cimento belonged to, 81; method of, disproved by Jungius, 168; intrigue against *Journal des sçavans*, 200; *Ratio studiorum* of, 215; successful experimenters, 218; dominant at University of Paris, 224; controlled the Catholic German universities, 227; utensils of, envied by Oxford student, 246

Jourdain, Charles, *Histoire de l'Université de Paris*, 220

Journal, the scientific, 198–209; see Scientific journals

Journal des sçavans, founded by Denis de Sallo, 56, 199; recorded the main occurrences in the Académie des Sciences, 153, 200, 201; purpose of, 200–202; under Gallois and La Roque

200; contents of, 201–2; model for periodicals, 202; only rival of, 207; copies of, 208

Journaux de médecine, published by La Roque, 203

Jungius, Joachim: on nomenclature in botany, 12; founded the Societas Ereunetica, 166, 167; on teaching in German, 167, 258 (n. 5); scientific attainments of, 167; purpose of the society, 168–69; Goethe on, 169

Keller, Dr. Ludwig, on Leibniz, 178

Kepler: on dynamics, 6; on lenses, 7; basis of laws of, 9; on the eye, 14–15; on the use of infinitesimals, 18; a mathematical physicist, 33–34; never saw a Kepler telescope, 67; not affiliated with a university, 257

Kiel, University of, freedom of thought promulgated at, 256

King, Dr., transfusion of blood by, 119

Kirch, to be in charge of Berlin Observatory, 191

Kircher, Athanasius: the *Arca Noe* of, 12; articles by, in *Journal*, 201; gave course in experimental physics at Würzburg, 230

Königsberg, lectures in vernacular at, 232 (n. 74)

Kyper, Albert, in charge of the clinic at Leyden, 252 (n. 154)

L'Ecluse: on classification of flowers, 12; traveled with the Fuggers, 64–65

L'Hôpital, favored the Newtonian calculus, 157

La Hire, Philippe de: and Picard, made map of France, 156; member of the Académie, 156; opposed the Newtonian calculus, 157

La Roque, Abbé: editor of *Journal des sçavans*, 200; and of *Journaux de médecine*, 203

Laboratories, scientific: the creation of seventeenth century, 4–6; "riches" of Bacon's "House of Salomon" a series of, 43; of Medici, 55, 77–78; of Cesi, 55; of Duke of Orleans, 56; of Charles II, 57; of Prince Rupert, 57; of Boyle, 58–59; of Van Helmont, 64; societies necessary to supply the, 67; planned by Accademia dei Lincei, 75; of the Cimento, 78; of Philosophical Society of Oxford, 98–99; of Royal Society, 103, 111–16; of Académie des Sciences, 147, 148, 159, 162; of Collegium Curiosum at Professor Sturm's home, 175; of Academy of Berlin, 192; foundation of, advised by Leibniz for Russia, 196; for Vienna, 196; necessity for, 216; at University of Bologna, 219, 254; at University of Altdorf, 231–32, 254; lack of, at Oxford, 246; establishment of, at Oxford, 247; at University of Leyden, 252, 254; none at Erlangen, 259; developed by the scientific societies, 260; of Bacon's "House of Salomon," 264–70; *see* Experimentation, Instruments

Latin: barrier to many, 54, 214; language of educated Germans, 165; colloquial language of the universities, 214, 220, 221–22, 237, 238; opposed by Descartes, 45; Jungius, 167; Leibniz, 181–82; Thomasius, 232–33; Webster, 243; use of, in universities must be changed, 216; was retained, 255; *see* Scholasticism, Vernacular

Laud, Abp., promulgated statutes at Oxford, 237

Le Boe [Sylivus], Francis de: professor of medicine at Leyden, 252; L. Schacht on method of teaching of, 252 (n. 154)

Leclerc, editor of *Bibliothèque universelle et historique*, 208

Leibniz, Gottfried Wilhelm von: on the organic origin of fossils, 17; on higher plane curves, 18; on the calculus, 18; librarian at court of Hannover, 65; on knowledge won from nature and experience, 66; co-operated with societies, 68; champion of learned societies in Germany, 166, 167; articles by, in the *Miscellanea curiosa*, 172; the Berlin Academy due to, 177; Frederick the Great on, 178; Keller on, 178; Harnack on, 178, 258 (n. 5); Kuno Fischer on, 178; scope of interests of, 178–79; his *Lehrjahre*, 179–81; objected to existing educational system, 181–82; to Latin, 181–82; to universities, 182; two great ideas of, 182; first project of, for learned society, 183–84; plans of, for learned

INDEX 299

societies, outlined in two programs, 184–86; as inventor, 186; sojourn of, in Paris, 186–87; F.R.S., 187; calls to learned men to found an academy, 187; purpose of the society, 187–88; librarian at court of Hannover, 188; other societies planned, 189; journey of, through Germany and Italy, 189; plans focused on Berlin, 189; the Electress planned an observatory at Berlin, 190; Jablonski aided in planning the Berlin Academy, 191; becomes president of Berlin Academy, 191; labored ten years in establishing, 192; friction with Jablonski, 193; unpopularity of, 193; articles by, in first volume of the *Miscellanea*, 194; denied his salary, 194; a potential founder of Dresden Academy, 195; efforts with Peter the Great, and in Vienna, 196; death of, 196; quarrel of, with Newton, 199; article by, in *Journal*, 201; not affiliated with a university, 257

Leipzig, University of: work required of professor at, 229; Christian Thomasius starts reform at, 233; no change in organization of, 259

Lémery, Nicholas: appointed to the Académie, 158; reports on chemistries of, in *Journal*, 201; not affiliated with a university, 257

Lenses, knowledge of, in 1600 and in 1700, 7

Leurechon's *Récréation mathématique*, editions of, 65

Leuwenhoeck: microscopic researches of, 13, 14; and his microscopes, 64; papers and letters of, in archives of Royal Society, 130; bequeaths microscopes to Royal Society, 130; article by, on the generation of man, in Bayle's *Nouvelles*, 207–8; other publications of, 208; not affiliated with a university, 257

Lexis, W., on the conservatism of the universities, 261

Leyden, University of: Huguenots, Puritans, and Germans at the, 251; Calvinistic doctrine prescribed at, 251; clinical teaching at, 252–53, 255; greatest students of medicine at, 252–53; Cartesian teaching at, 253; established its own laboratory, 254

Lhwyd, on fossils, 16

Light: theories of, 7; experiments of the Cimento on velocity of, 88; Huygens' undulatory theory of, 154; velocity of, calculated by Roemer, 154

Light and colors, Newton's theory of, presented to Royal Society, 135–36

Lightness, existence of positive, refuted, 87

Linnaeus, revolutionary work of, 13

Litmus paper, the properties of, experimented on, 88

Locke, John, pupil of P. Stahl, 99 (n. 29)

Logarithms, evolution of, 18

Logic, study of: should be minimized, 216; hardly lessened, 254

London Royal Society; *see* Royal Society

Louis XIV: Persons touched by, 20; financed the Académie des Sciences, 139; planned reforms for the universities, 224 (n. 34)

Louvain, University of, 253

Louvois, François Michel le Tellier, Marquis de: succeeded Colbert, 156; demands *recherche utile* of the Académie, 157

Louvre, the Académie des Sciences assigned rooms in the, 161

Lower, Dr. Richard: pupil of P. Stahl, 99; transfusion of blood by, 118, 119

Lymphatic system, discovered, 14

Lynceographum, the, quoted, 75

Lynx, a, the device of the Lincei, 74

Machines and mechanisms, catalogue of, by Binot and Couplet, 155

Magalotti, Lorenzo: secretary of the Cimento, 80; published the *Saggi*, 83; sent a copy to the Royal Society, 88; on *Musaeum* of Royal Society, 115

Magiae naturalis, the, of Della Porta, 73–74

Magnetism: knowledge of, in 1600 and in 1700, 8; Gilbert's work on, 22–23; experiments on, by the Cimento, 87

Magneto-mathematical society, a, planned by Leibniz, 189

Malpighi: microscopic researches of, 13; on circulation of the blood, 14;

on the capillaries, 38; correspondence with, 124; articles on, in *Journal des sçavans*, 201; chair of medicine founded in Bologna for, 219

Marburg, University of: Johann Hartmann professor of chemistry at, 230; Papin professor of physics at, 230

Marcgrave, George, astronomer, geographer, and naturalist, 64

Mariotte: on hydrostatics, 7; on blind spot in the eye, 119; studied the flow of sap, 151; experiments of, 154; discovered blind spot of the eye, 154; interested in mechanics, 155; planning the Versailles cascades, 156; opportunity given to, 163; contributed to the *Journal*, 201; not affiliated with a university, 257

Marsiglio, Ludovico Ferdinando, Count: a great experimenter, 55; bequeathed his laboratories to University of Bologna, 219

Marsili, Alesandro, a member of the Cimento, 80

Massari, B., organized *Chorus anatomicus* and carried on dissections at Bologna, 219

Mästlin, Michael, afraid to teach Copernican doctrine at Protestant Tübingen, 255

Mathematics: knowledge of, in 1600 and in 1700, 17–18; the supreme instrument of research in theoretical physics and astronomy, 18; the explanation of Euclid, 214; no, at Cambridge for thirty years, 241

Maurylocus, studied lenses, 7

Mechanisms, the Académie des Sciences ordered to interest themselves in, 155

Medical knowledge in 1600 and in 1700, 13–15

Medical School of Florence, vivisection used at, 80

Medici, Ferdinand and Leopold: amateurs in science, 55–56; founders of the Accademia del Cimento, 77; instruments and collections of, 77; observations and experiments of, 78; Leopold, the head of the Cimento, 78; hatching of eggs, the hobby of Leopold, 80; alcohol thermometer of Leopold, 83; hygrometer devised by Ferdinand, 84; Leopold a correspondent of the Royal Society, 104

Medicina practica: flourished in Germany, 185; Leibniz on, 186

Medicine, faculty of: represents scientific interest, 215; study of, must be reformed, 216; experimental, in Italy at Padua, 218; 255; at University of Paris, 220–21; at Oxford, 236; at London, 255; *see* Anatomy, Dissection

Mercure sçavant, published by Nicholas de Blegny and Gautier, 203

Merret, Dr.: at earliest meetings of Royal Society, 93; friend of Harvey, 95

Mersenne, Morin: on acoustics, 7, 88; work of, 140–41; scientists who gathered about, nucleus of Académie des Sciences, 141–42; translated works of Galileo, 140; popular writings of, 140; Cassini on meetings at cell of, 142, voluminous correspondence of, 140; 198

Messena, scientists of the Cimento on faculties of, 219

Metaphysics: emphasis on, should be minimized, 216; hardly lessened, 254

Micro-organisms, the study of, 116–17

Microscope, the: created a new world, 13–14, 21, 22; Galileo's first used in research, 26; Hooke and Sprat on, 53; made by Galileo for the Lincei, 76; named by a member 76

Microscopes of Leuwenhoeck, cabinet of, given to Royal Society, 130

Mineralogy, knowledge of, in 1600 and 1700, 15

Miscellanea curiosa: published, 171, 203; Oldenburg on the, 172; German sentimentalism in, 172; Prutz on the, 173–74; praised by Bayle, 174; extracts from, in *Journal*, 202; *see* Collegium Naturae Curiosorum

Miscellanea naturae curiosorum; see Miscellanea curiosa.

Molineux, William, inventor of hygroscope, 63

Montaigne, held university system valueless, 258

Montmort, Hubert de, meetings of scientists at house of, 143

INDEX

Montpellier, University of, no account of, found, 220 (n. 18)

Moray, Sir Robert: an amateur of science, 61; "soul of the meetings" of Royal Society, 102; on barnacles, 104; interested Charles II in Royal Society, 104; charter member, 105; on worms in vinegar, 117

Morin, Jean Baptiste, opposed the Copernican system, 223

Morison, English botanist, supervisor of gardens at Blois, 56

Moritz, Johann: a science-loving Dutch amateur, 63; Swainson on, 63; expedition to Brazil of, 64

Musaeum of the Royal Society, 114

Musschenbroek, P. von, translated the *Saggi* of the Cimento into Latin, 88–89

"Naissance" of independence from classical thought, 50

Napier, on logarithms, 18

Neil, Sir Paul, charter member of Royal Society, 100, 105

Newton, Isaac: optical researches of, 5; on dynamics, 6; on light, 7; law of attraction of, 9; on higher-plane curves, 18; on the calculus, 18; and the Christian epic, 20; co-operated with societies, 68; gave telescope to Royal Society, 114; made member of Royal Society, 132, 134–35; close connection of Royal Society with, 134–38; reflecting telescope of, 134, 201; communicates theory of light and colors, 135–36, 257 (n. 2); criticized by Hooke and Huygens, 136, 201; publication of the *Principia* and quarrel with Hooke, 137–38; quarrel of, with Leibniz, 199; a tireless experimenter, 248; Lucasian professor at Cambridge, 248–49; philosophy of, introduced, 249, 254; on the scientific society, 260 (n. 11)

Nouveau journal des savants dressé à Rotterdam, edited by Étienne Chauvin, 208

Nouvelles de la république des lettres: of Pierre Bayle, 207–8; contents of, 207–8; copies of, 208

Nouvelles descouvertes sur toutes les parties de la médecine, published by Nicholas de Blegny, a charlatan, 203

Observatory, modern: creation of seventeenth century, 5; of Hevelius, 64; of Greenwich, and of Flamsteed, 133–34; of Paris established, 152, 162; Leibniz on plan of Electress for an, at Berlin, 190; of Berlin Academy, 191, 193, 194; necessity for, 216; at University of Altdorf, 230; national, created by scientific societies, 260; at Bacon's "House of Salomon," 264–65

Oldenburg, Henry: correspondent of the Cimento, 82; charter member and secretary of Royal Society, 105; on a college for the Royal Society, 122; extensive correspondence of, 123–24; read articles in scientific journals, 124; published the *Philosophical Transactions*, 124–25; on securing inventions, 130; visit of, to Mersenne, 142; on the *Miscellanea curiosa*, 172

Olivia, Antonio, member of the Cimento, committed suicide, 80

Optics: knowledge of, in 1600 and in 1700, 7; Kepler's work in, 33; problems of, studied, 117; at Cambridge, 248

Orleans, Duke of, an amateur in science, 56

Orthodoxy, religious: of scientists, 20; of Descartes, 46; of Boyle, 57–58; of *Acta eruditorum*, 204–5; must be removed, 216; not demanded in Italian universities, 217, 255; demanded at University of Paris, 220; 221–23; emphasis on, in German universities, 227–28; at universities of Holland, 251

Otiosi, the: 74; *see* Della Porta

Oxford: meetings of members of Royal Society at, 95–96; an Invisible College at, 95 (n. 12), 241, 244; condition of chemical laboratory, 242

Oxford Philosophical Society, the, 99 (n. 32)

Oxford University: Halley Savilian professor at, 134; science at lowest ebb in, 236; Regius professorship, 236; disputations in medicine, 236; Savile professorships endowed, 236; Laudian statutes, 237–38; Latin used exclusively, 238; Wallis Savilian professor at, 241; remarkable set of men at, 241; no science at, 242, 245; students of, interested in science, 246; Ashmolean Museum at, 247; science little cultivated at, 247; no change in statutes of, 259

Padua, University of: under control of Venice, 217; first botanical garden and anatomical theater at, 218; Vesalius at, 218; scientists of the Cimento at, 219

Paleontology, knowledge of, in 1600 and in 1700, 15–16

Palmer, Dudley, charter member of Royal Society, 105

Papin, Denis: gave boiler to Royal Society, 114; experiments of, on circulation of the blood, 207; taught physics in Marburg, 230

Paracelsus: on chemistry, 10–11; the iatrochemists followers of, 15; lectured in German in Basel, 232 (n. 74); held university system valueless, 258

Paris, University of: most important in Europe, 220; only Catholics admitted, 220; statutes and courses, 220–21; stagnation at, 221; liberty of thought suppressed, 221, 223; opposition to teaching of Descartes, 222–23; intolerant in religion and education, 223–24; battle of, against Jansenism and the Jesuits, 224; botanical garden and polyclinics at, 225; Faculty of Arts at, placed Descartes beside Aristotle, 256

Pascal, Blaise: on hydrostatics, 7, 33; on the theory of probability, 18; on atmospheric pressure, 33; associate of Mersenne, 141; meetings of scientists at house of, 142; quarrel of, with Torricelli, 199; not affiliated with a university, 257

Paul V, Pope, friendly to Galileo, 28

Pecquet, Jean: on the lymphatic system, 14; discoverer of the thoracic duct, 38; human dissections by, 153; opportunity given to, 163; medical discoveries of, in *Journal des sçavans*, 201

Peiresc: a famous amateur of science, 56; member of the Lincei, 74; won Gassendi to study of Galileo and Kepler, 142; voluminous correspondence of, 198

Pendulum, the isochronous: of Huygens, 18; tested, 119

Pendulum clock, Huygens,' 154, 162

Pensionnaires, of the Académie des Sciences, 148, 160

Périer, on atmospheric pressure, 33

Perrault, Charles: interested Colbert in science, 144, 146; dissection of animals by, 149–50; the *Histoire des animaux* of, translated, 150; joint author of *Histoire des plantes*, 150; botanical questions by, 150–51; studied the flow of sap, 151; human dissections by, 153, 201; mechanisms by, in catalogue, 155

Peter the Great, gave pension to Leibniz, 196

Petty, Sir William: on mortality tables, 18; an untiring experimenter, 62; at Oxford, 95 (n. 12); work assigned to, 103–4; charter member of Royal Society, 105; ship of, wrecked, 120

Philosophers, the: of the seventeenth century, 22; Sir Francis Bacon, 39–44; Descartes, 44–50

Philosophi Inquieti, an academy of experimenters at Bologna, 219

Philosophical Society of Oxford, Sprat on the meetings of, 98; Anthony Wood on the, 98–99

Philosophical Transactions: publication of, begun by Oldenburg, 124, 202; Huxley on, 125; Oldenburg on purpose of, 126–27; contents of first number 127; battle-ground of scientific opinions, 127; essential feature of, 128; published in Latin abroad, 128; extracts from, in *Journal des sçavans*, 202; standard for publications of scientific societies, 202

Philosophy, faculty of: reform needed in studies under, 215, 216; little change in, 254

Phosphorescence, studied by the Cimento, 88

Physics: before the seventeenth century, Rosenberger on, 3–4; knowl-

INDEX

edge of, in 1600 and in 1700, 6–7; progress in, 8, 19; pioneers of, 21–34, 50–52; laboratory of, of Professor Sturm, 175, 230; must be established in universities as separate subject, 216; experimental, taught in Italian universities, 218; Aristotle's upheld, 223, 224; often taught by professor of medicine, 230; experimental, at University of Würzburg, 230; Bacon regrets lack of, in universities, 239 (n. 105); established at universities, 254; *see* Experimentation, Instruments, Laboratory

Physiology, knowledge of, in 1600 and in 1700, 13–14

Picard: asked Louis XIV for an observatory, 151; expedition of, to Uranienburg, 152; perfecting the telescope, 153; made map of France, with La Hire, 156

Pisa, University of: under control of Florence, 217; scientists of the Cimento in faculties of, 219

Pliny, errors in, refuted by Renaldini, 80–81

Plot, Dr., professor of chemistry at Oxford, 247

Pneumatics, knowledge of, in 1600 and in 1700, 7

Poggendorff, on the work of the Cimento, 89

Polyclinics, established at University of Paris, 225, 255

Pontchartrain, Louis Phelypeaux, Comte de, succeeded Louvois, 159

Porphyry, *Isagoge* of, studied at Dublin University, 238

Potter, Francis, on transfusion of blood, 62

Principia, Newton's, publication of, 136, 137; *see* Newton

Prizes, for solution of problems, origin of, 140

Progress in the different sciences in the seventeenth century, 5–19, 21; how spread, 52–54; what Cimento did for, 89; what Royal Society did for, 138; what the Académie des Sciences did for, 162–64; conditions requisite for, in the universities, 215–16; how far fulfilled, 253–56; active centers of, 253; fostered in scientific societies, 259

Public lectures and meetings of the Royal Society established, 123

Quarrels regarding scientific discoveries, 199

Ramus, Peter: taught at Collège de France, 225–26; book of, interdicted at Freiburg, 228; held university system valueless, 258

Rarities, in *Musaeum* of Royal Society, Magalotti on the, 115

Ratke, pedagogic reformer, 167

Ray, John: on classification of plants and animals, 13; member of the Royal Society, 129; lost position at Cambridge, 248

Redi, Francesco: member of the Cimento, founded Medical School of Florence, 80; denied spontaneous generation, 80

Reforms, fundamental: needed in the universities, 215–16; how far begun, 253–56

Regiomontanus, a German, 185

Register of Oxford, the, 235 (n. 82), 236

Regius (Van Roy), Henry, taught Cartesianism at Utrecht, 253

Regius professorship at Oxford endowed by Henry VIII, 236

Renaldini, Carlo, member of the Cimento, refuted errors in Aristotle, Pliny, and Gilbert, 80–81

Renaudot, Theophraste, established a polyclinic, 225 (n. 36)

Renery, taught Cartesianism at Utrecht, 253

Ricci, Michael Angiolo: corresponding member of Cimento, 81; editor of *Giornale de' literatti*, 81, 202

Riccioli: opponent of the Copernican system, 89; Jesuit, skilful experimenter, 218

Richer, on variation of length of the second pendulum at Cayenne, 152

Roberval: on cycloid curves, 18; on the use of infinitesimals, 18; corresponded with Descartes, 140; associate of Mersenne, 141; on free fall, and periodicity of the comet, 154; interested in mechanics, 155; studying games of chance, 156; balance of, described in *Journal*, 201; at Collège de France, 226

Roemer: Danish astronomer, brought to Paris, 152; calculated the velocity of light, 154; interested in mechanics, 155; lost to the Académie, 156

Rohault, J., *Physics* of, translated into Latin by S. Clarke with Newton's views as notes, 249

Rolfink, Würzburg professor of anatomy, 232

Rolle, opposed Newtonian calculus, 157

Rooke, Laurence: Oxford member of Royal Society, 96, 241; Aubrey on, 97; professor at Gresham College, 100; Society met at chambers of, 101

Rosenberger: quoted, 3–4, 21; on the members of the Cimento, 89–90

Rosicrucians, Leibniz connected with the, 179, 180

Rostock: Societas Ereunetica at, 167; lectures in Low German at, 232 (n. 74)

Royal Society, the, 68, 91–138; an informal organization, 91; Sprat on, 92–93; Wallis on beginning of, 93; the original members, 93–95; business of the London meetings, 95; meetings at Wadham College, Oxford, 95–96; the Oxford members, 96–98; Stahl's laboratory, 99; removal to, and meetings at Gresham College, 99–100; organized, 101; first *Journal* book and weekly meetings, 101–2; formative period, 103–4; charter granted by Charles II, 104–5; Fellows appointed, 107; members admitted, 107 (n. 59); statutes adopted, 107–10; personnel of the Fellows, 110–11; a laboratory planned, 112; the instruments and *Musaeum*, 112–14; work of, 115–17; under guidance of curator Hooke, 117; study of microorganisms, 117; dynamics and optics, 117; air in breathing and burning, 117–18; transfusion of blood, 118–19; testing experiments of others, 119–20; Sprat on the practical investigations of, 120–21; question of forming a college, 122; public lectures and meetings established, 123; correspondence with scientists, 123–4; examination of scientific journals, 124; *Philosophical Transactions* begun, 125–27; character of their contents, 127–28; publication of scientific works by, 129–30; encouraged research work, 130; defenders and opponents of, 131–32; famous members of second decade of, 132–34; close connection of, with the works of Newton, 134–35; first of reforming bodies, 138; Leibniz a Fellow of, 187; opinions of scientists on work of, 260 (n. 11)

Rupert, Prince: an amateur of science, 57; inventions of, 57

Sachs, J., on the botanical knowledge of 1600, 12

Sachs von Lowenhaimb, Dr. Philipp Jacob: enlarged and reorganized the Collegium Naturae Curiosorum, 170–71; paper by, published, 171–72, 201

Sacri Romani Imperatoris Academia Naturae Curiosorum, 173

Saggi di naturali esperienza fatte nell Accademia del Cimento: analysis of, 83–88; translated into English, Latin, and French, 88–89; republished in the *Atti e memorie*, with a supplement, 89; experiments recorded in, tested by committee of the Royal Society, 120; translated at instance of the Royal Society, 129

St. Andrews, University of, 235; Newton's ideas introduced in, by David Gregory, 250

St. Simon, on the Duke of Orleans as a lover of science, 56

Sallo, Denys de: founder of *Journal des sçavans*, 56, 199–200; lost license through Jesuit criticism, 200

Salzburg, University of, controlled by Benedictines, 227

Sanctorius, taught in Padua, 219

Saturn and his rings: observations on, 89; oval shape of, 95; Huygens on the rings of, 143

Savile, Sir Henry: charter member of Royal Society, 105; endowed professorship at Oxford, 236

Schacht, Lucas, on Le Boe's method of clinical instruction at Leyden, 252 (n. 154)

Scheiner, Jesuit, contest of, with Galileo, 27

Scheuchzer, on fossils, 16
Scholasticism: had created a caste, 54; experimental science in competition with, 54; Leibniz opposed to, 181; blow dealt at, by Thomasius, 234; see Aristotle, Conservatism, Vernacular, Universities
School reform; see Education
Schott, Gaspard: first published Guericke's experiments with air pump, 51; gave course in physics at Würzburg, 230
Schrevelius, E., and Otto Van Heurne inaugurated clinical teaching at Leyden, 252 (n. 154)
Schwenter, popular work on experiments by, 66–67
Science, amateurs in: see Amateurs
Science, a faculty of, must supplement that of philosophy, 216
Science and religion, Boyle on reconciliation of, 58
Science in the universities; see Universities, science in the
Sciences: thrive only in a non-ecclesiastical atmosphere, 215; proved by facts, 253–54; must be studied as independent disciplines, 216; a tendency toward specialization in, 254; organized support of, derived from scientific societies, 261
Scientific advance, general, in the seventeenth century, 3–20
Scientific journals, the, 198–209; functions of the scientific societies, 198; unreliability of private correspondence, 199; Denis de Sallo starts the *Journal des sçavans*, 199; editors and contents of the early years, 200–201; authors of scientific articles, 201; *Philosophical Transactions*, 202; *Giornale de' litterati 'i Roma*, 202; *Miscellanea naturae curiosorum*, and the *Acta medica et philosophica hafniensia*, 203; *Nouvelles descourvertes de la médecine*, and *Mercure sçavant*, 203; *Journaux de médecine, Collectanea medico-physica*, 203; *Acta eruditorum* at Leipzig, 203–4; edited by Otto Mencke, included scholasticism and the new knowledge, 204; contents of the volumes, 205–7; Bayle's *Nou-velles de la république des lettres*, 207–8; *Histoire des ouvrages des savants*, and *Bibliothèque universelle et historique*, 208; *Dépêche du Parnasse*, and *Nouveau journal des savants*, 208

Scientific societies: aid rendered by the, xi; origin and need of, 67; enumeration of, 68–69; Italian, 73–88; German, 165–97; work of, for science, 259–63; have revolutionized the universities, 263; see Royal Society, Académie des Sciences

Scientists, individual, rôle of, 21–69
Scientists, many of the greatest, not affiliated with universities, 257, 259
Second-pendulum, variation in vibration of, 152
Sedley, William, endowed lectureship at Oxford, 237
Segni, first secretary of the Cimento, 80
Seventeenth century, the: a period of great significance, 3; scientific advance in, 3–20; introduced experimentation into science, 5, 19; "mutation" and elaboration in, 21; the scientists and the philosophers in, 22–52
Sidereus nuntius, the, of Galileo, 26–27
Sienna, the Academia Physico-critica in, 68–69
Sinclair, George, gave telescope to University of Glasgow, 250
Slingesby, Henry, charter member of Royal Society, 100, 105
Snellius, on refraction, 7
Societas Ereunetica, 166; established by Jungius, 167; its purpose, 168; its motto, 168; short-lived, 168
Societas Eruditorum Germaniae, planned by Leibniz, 183–84
Societas Pythagorea, established by Weigel, 180
Societas Theophilorum vel Amoris Divini, planned by Leibniz, 188
Spanheim, letters of Leibniz to, on an electoral society, 189
Spinoza: Leibniz a correspondent of, 179; offered a chair in the University of Heidelberg, 228 (n. 47)

Sprachgesellschaften, the, prepared the way for learned societies in Germany, 166

Sprat, Thomas: on the microscope, 53; on experimental science, 55; on science among the aristocracy, 57; on the genius of experimenting, 63; on the Royal Society, 92; on the meetings at Gresham College, 100; F.R.S., 107; experimenting by the Fellows, 115–16; on investigations of practical subjects, 120; defense of the Royal Society, 131; loyalty of, to university system, 258 (n. 5)

Stahl, Peter: taught chemistry at Oxford, 98, 245; became operator of the Royal Society, 99

Statutes of the reorganized Académie des Sciences, 160–61

Statutes of the Royal Society, extracts from, showing aim and methods, 108–10

Steam engine, inventor of first, 61

Stelluti, Francesco: first used microscope of Galileo for research, 26; member of the Lincei, 74; first used microscope in study of zoölogy, 76; work of, on bees published by the Lincei, 76

Steno, Nicolaus: an iatrophysicist, 15; on crystals, 15; on the earth's crust, 17; member of the Cimento, physician of the Medici, Protestant and Catholic, 81; experiments of, tested, 119–20; performed dissections before French scientists, 143, 144; associate member of Académie des Sciences, 156; studied at Leyden, 253

Strassburg, University of, relative attendance in faculties of, 228 (n. 50)

Stubbe, Dr., attack of, on Royal Society, 132

Sturm, Christopher: a guest of the Cimento, 82; on Borelli's *De motu animalium*, 172; at University of Altdorf, 175, 230; established the Collegium Curiosum sive Experimentale, 175–76; showed experiments publicly, 176

Swainson, on Johann Moritz, 63–64

Swammerdam, microscopical observations of, 13, 14; student at Leyden, 253; not affiliated with a university, 257

Sylvius: *see* La Boe, Francis de

Tannery, Paul, on scientific societies, 261–62

Telescope, first: claimed for Della Porta, 7; results of invention of, 21, 22; Galileo's, the first to be pointed to the heavens, 26; Aristotle and the, 29 (n. 18); Kepler's, 33–34; Glanvill on the, 53; reflecting, of Newton, 134; perfected by Huygens and others, 153; Fontenelle on, 162

Telesio, held university system valueless, 258

Theology: matters of, excluded in Royal Society, 95; faculty of, most important, 213; must take second place, 216; dominant in German universities, 227; retained pre-eminence, 255; *see* Orthodoxy

Thermometers: Grand Duke Leopold interested in, 77; studied by the Cimento, 83–84; the best, made by Italians, 90; first used for medicine, 219

Thesaurus Mexicanus, most ambitious work of the Lincei, 76

Thevenot, Melchisedec: correspondent of the Cimento, 82; scientists met at house of, 144

Thomasius, Christian: on necessity for witch trials, 20; hero of German university reform, 232; lectured in German at Leipzig, 233; attack of, on Latin, 233–34; on the university system, 234, 258 (n. 5); foundation of University of Halle, 234; *see* Vernacular

Thomson, Thomas, on work of Royal Society, 116

Torricelli: on atmospheric pressure, 7; on hydrostatics, 7, 33; discovered the barometer, 33; pupils of, in the Cimento, 77; the Medici brothers, 77; work of, studied by Royal Society, 92; quarrel of, with Pascal, 199

Touch, belief in efficacy of the, 20

Tozzetti, Gio Targione, published the *Atti e memorie del Accademia del Cimento*, 89

INDEX 307

Trew, had observatory at Altdorf, 230

Tschirnhausen, Duke: devoted to science, 65; associate member of Académie des Sciences, 156; opposed the Newtonian calculus, 157; planned learned society in Saxony, 195

Tübingen, University of: religious orthodoxy required at, 227; Mästlin afraid to teach Copernican doctrine at, 255

Tuke, on the early meetings of French scientists, 143

Tycho Brahe: on astronomy, 8–9; Kepler's laws based on observations of, 9; error of, corrected by Picard, 152

Uffenbach, on condition of Royal Society's laboratory at Oxford, 242

Universities: contrast of system of, with experimental science, 54–55; Leibniz, opponent of, 178, 180–82; courses leading to degree at, 213; faculties of, 214; methods of, 214; opposition to freedom of thought, 215; fundamental reforms needed, 215–16; of Italy, 217–19; of France, 220–26; of Paris, 220–24; stagnation of, at Paris, 221; opposition to Descartes, 222–23; intolerance in religion and education, 224; Collège de France, 225–26; of Germany, 226–35; Protestant and Catholic, 227–28; Thomasius' opposition to, 232–35; of England, 235–50; reactionary Laudian statutes, 237–38; Bacon on, 239–41; flood of criticism against, 242–45; how Newtonian philosophy took root in, 249–50; of Holland, 250–56; progress made in the seventeenth century, 253–56; contributed little to advancement of science, 257, 258–59; many scientists and thinkers not affiliated with the, 257–59; conservatism of, severely criticized, 258; work of the scientific societies, 259–60; Harnack on universities and academies, 261; Lexis on, 261; Tannery on, 261–62; Eulenberg on, 262; societies the *Kulturträger*, 262; revolution in the universities, 263; see Aristotle, Conservatism, Scholasticism

Universities, the: science in, 213–56; traces of, 215; of Italy, 217–19; of Paris, 220, 224–25; of Germany, 229–32; of England, 235–38; Bacon on, 239–41; lack of, criticized, and reply, 242–45; Oxford students interested in, 246–47; of Cambridge, 247–50; spreading in England, 249–50; of Holland, 251–53; progress of, in the seventeenth century, 253–55; contributed little to advancement of knowledge, 257; see Aristotle, Conservatism

Uranienburg, scientific expedition to, 152

Utrecht, University of: had German and English students, 251; Calvinistic doctrine prescribed at, 251; Cartesian teaching at, 253

Vacuum experiments of the Cimento, 85

Valerius, Lucas, resigned from the Lincei, 76

Van Helmont, J. B.: an iatrochemist, 15; a pioneer observer, 38; an amateur of science, 64; experiments of, tested, 119–20; not affiliated with a university, 257

Van Heurne, Otto, and E. Schrevelius inaugurated clinical teaching at Leyden, 252 (n. 154)

Varignon, favored the Newtonian calculus, 157

Venice, learned society in, 68

Vernacular, the: Galileo's work in, 28, 31; Descartes' *Méthode* in, 45; language of science, 54; low development of, in Germany, 165; Jungius pleads for, 167; Leibniz advocate of, 178–79; 181–82, 188; Elector advocate of, in Berlin Academy, 191; must become vehicle of university teaching, 216; prohibited at Paris, 221; sporadically used at German universities, 232 (n. 74); Christian Thomasius, champion of, 232–33; forbidden at English universities, 238; want of, criticized, 243; not introduced in universities, 255

Vesalius: founder of non-Galenic science of anatomy, 34; at University of Padua, 218

Vichy waters chemically examined, 154

Vienna Academy, plan for, by Leibniz, defeated by the Jesuits, 196

Viviani, Vincenzo: pupils of, in the Cimento, 76; the Medici brothers, 77; member of the Cimento, 78–79; last pupil of Galileo, 79; constructed first barometer, 79; reconstructed works of Apollonius, 79; associate member of the Académie des Sciences, 156

Vivisection, practiced at medical school of Florence, 80

Voetius, rector at Utrecht, opposed Harvey and Descartes, 253

Wadham College, scientific interest at, 96

Waller, Richard, translated the *Saggi* of the Cimento, 88

Wallis, John: on the use of infinitesimals, 18; charter member of Royal Society, 93, 105; on its earliest meetings, 93; a great mathematician, 93; on the work of the Society, 95; on the meetings in Oxford, 95–96; pupil of Peter Stahl, 99; F.R.S., 107; on transfusion of blood, 118; voluminous correspondence of, 199; on dearth of mathematics at Cambridge, 241; Savilian professor at Oxford, 241

Ward, Seth: Oxford member of Royal Society, 96; expelled from Alma Mater, Cambridge, 241; reply of, to Webster, 244–45

Webster, J.: *Academiarum examen* of, Mullinger on the, 243–44; reply to, by Wilkins and Ward, 244–45

Weigel, Erhard: Leibniz studied mathematics with, 180; plan of, for a monopoly of calendar reform, 190; made an example of intolerance, 228

Whiston, Lucasian professor at Cambridge, 249–50

Wilkins, John: charter member of Royal Society, 93, 94, 105; Oxford meetings at lodgings of, 96, 241; chairman of meetings, 101; secretary of Royal Society, 106; reply of, to Webster, 244; preached at, from the Oxford pulpit, 246

Willis, Dr. Thomas: on brain and nerves, 14; a foremost English physician, 96, 97–98, 241; studied at Leyden, 253

Willoughby, Francis: on classification of plants and animals, 13; amateur of science, 61; F.R.S., 107; *History of Fishes* of, published by John Ray and the Royal Society, 129; publications of, in French journals, 208; not affiliated with a university, 257

Winthrop, John: made F.R.S., 107; on building ships in America, 121

Witchcraft: belief in, 20; Della Porta accused of meddling with, 90

Wood, Anthony, on "The Royall Societe at Oxon," 98

Woodward, on fossils, 16

Worcester, Marquis of, an amateur of science, 61

Worcester Hall, Oxford University, scientific courses in, 247

Wren, Christopher: on motion of celestial bodies, 9; on the use of infinitesimals, 18; Oxford and charter member of Royal Society, 96, 105, 241; scientific versatility of, 97; pupil of Peter Stahl, 99; professor at Gresham College, 100; Royal Society organized after a lecture by, 101; F.R.S., 107; plans of, for a college for the Royal Society, 122 (n. 127)

Wren, Mathew, charter member of Royal Society, 105

Würzburg, University of: required a *confessio fidei*, 228; relative attendance in faculties of, 228 (n. 50); experimental physics at, 230; anatomical theater at, 231

Zoölogy: knowledge of, in 1600 and in 1700, 12–13; papers on, in the *Miscellanea curiosa*, 201

HISTORY, PHILOSOPHY AND SOCIOLOGY OF SCIENCE

Classics, Staples and Precursors

An Arno Press Collection

Aliotta, [Antonio]. **The Idealistic Reaction Against Science.** 1914

Arago, [Dominique François Jean]. **Historical Eloge of James Watt.** 1839

Bavink, Bernhard. **The Natural Sciences.** 1932

Benjamin, Park. **A History of Electricity.** 1898

Bennett, Jesse Lee. **The Diffusion of Science.** 1942

[Bronfenbrenner], Ornstein, Martha. **The Role of Scientific Societies in the Seventeenth Century.** 1928

Bush, Vannevar. **Endless Horizons.** 1946

Campanella, Thomas. **The Defense of Galileo.** 1937

Carmichael, R. D. **The Logic of Discovery.** 1930

Caullery, Maurice. **French Science and its Principal Discoveries Since the Seventeenth Century.** [1934]

Caullery, Maurice. **Universities and Scientific Life in the United States.** 1922

Debates on the Decline of Science. 1975

de Beer, G. R. **Sir Hans Sloane and the British Museum.** 1953

Dissertations on the Progress of Knowledge. [1824]. 2 vols. in one

Euler, [Leonard]. **Letters of Euler.** 1833. 2 vols. in one

Flint, Robert. **Philosophy as Scientia Scientiarum and a History of Classifications of the Sciences.** 1904

Forke, Alfred. **The World-Conception of the Chinese.** 1925

Frank, Philipp. **Modern Science and its Philosophy.** 1949

The Freedom of Science. 1975

George, William H. **The Scientist in Action.** 1936

Goodfield, G. J. **The Growth of Scientific Physiology.** 1960

Graves, Robert Perceval. **Life of Sir William Rowan Hamilton.** 3 vols. 1882

Haldane, J. B. S. **Science and Everyday Life.** 1940

Hall, Daniel, et al. **The Frustration of Science.** 1935

Halley, Edmond. **Correspondence and Papers of Edmond Halley.** 1932

Jones, Bence. **The Royal Institution.** 1871

Kaplan, Norman. **Science and Society.** 1965

Levy, H. **The Universe of Science.** 1933

Marchant, James. **Alfred Russel Wallace.** 1916

McKie, Douglas and Niels H. de V. Heathcote. **The Discovery of Specific and Latent Heats.** 1935

Montagu, M. F. Ashley. **Studies and Essays in the History of Science and Learning.** [1944]

Morgan, John. **A Discourse Upon the Institution of Medical Schools in America.** 1765

Mottelay, Paul Fleury. **Bibliographical History of Electricity and Magnetism Chronologically Arranged.** 1922

Muir, M. M. Pattison. **A History of Chemical Theories and Laws.** 1907

National Council of American-Soviet Friendship. **Science in Soviet Russia: Papers Presented at Congress of American-Soviet Friendship.** 1944

Needham, Joseph. **A History of Embryology.** 1959

Needham, Joseph and Walter Pagel. **Background to Modern Science.** 1940

Osborn, Henry Fairfield. **From the Greeks to Darwin.** 1929

Partington, J[ames] R[iddick]. **Origins and Development of Applied Chemistry.** 1935

Polanyi, M[ichael]. **The Contempt of Freedom.** 1940

Priestley, Joseph. **Disquisitions Relating to Matter and Spirit.** 1777

Ray, John. **The Correspondence of John Ray.** 1848

Richet, Charles. **The Natural History of a Savant.** 1927

Schuster, Arthur. **The Progress of Physics During 33 Years (1875-1908).** 1911

Science, Internationalism and War. 1975

Selye, Hans. **From Dream to Discovery: On Being a Scientist.** 1964

Singer, Charles. **Studies in the History and Method of Science.** 1917/1921. 2 vols. in one

Smith, Edward. **The Life of Sir Joseph Banks.** 1911

Snow, A. J. **Matter and Gravity in Newton's Physical Philosophy.** 1926

Somerville, Mary. **On the Connexion of the Physical Sciences.** 1846

Thomson, J. J. **Recollections and Reflections.** 1936

Thomson, Thomas. **The History of Chemistry.** 1830/31

Underwood, E. Ashworth. **Science, Medicine and History.** 2 vols. 1953

Visher, Stephen Sargent. **Scientists Starred 1903-1943 in American Men of Science.** 1947

Von Humboldt, Alexander. **Views of Nature: Or Contemplations on the Sublime Phenomena of Creation.** 1850

Von Meyer, Ernst. **A History of Chemistry from Earliest Times to the Present Day.** 1891

Walker, Helen M. **Studies in the History of Statistical Method.** 1929

Watson, David Lindsay. **Scientists Are Human.** 1938

Weld, Charles Richard. **A History of the Royal Society.** 1848. 2 vols. in one

Wilson, George. **The Life of the Honorable Henry Cavendish.** 1851